The Theory of
Ecological Communities

MONOGRAPHS IN POPULATION BIOLOGY
EDITED BY SIMON A. LEVIN AND HENRY S. HORN

A complete series list follows the index.

The Theory of
Ecological Communities

MARK VELLEND

PRINCETON UNIVERSITY PRESS
Princeton and Oxford

Published by Princeton University Press, 41 William Street, Princeton, New Jersey 08540
In the United Kingdom: Princeton University Press, 6 Oxford Street, Woodstock,
Oxfordshire OX20 1TW

press.princeton.edu

Library of Congress Cataloging-in-Publication Data

Names: Vellend, Mark, 1973– , author.
Title: The theory of ecological communities / Mark Vellend.
Description: Princeton : Princeton University Press, 2016. I Series: Monographs in
    population biology I Includes bibliographical references and index.
Identifiers: LCCN 2015047633 I ISBN 9780691164847 (hardcover : alk. paper)
Subjects: LCSH: Biotic communities.
Classification: LCC QH541 .V425 2016 I DDC 577.8/2—dc23 LC record available
    at https://lccn.loc.gov/2015047633

British Library Cataloging-in-Publication Data is available

This book has been composed in Times LT Standard

Printed on acid-free paper. ∞

Printed in the United States of America

10 9 8 7 6 5 4 3

Contents

Acknowledgments

The ideas in this book, and the inspiration and encouragement to develop them, have come from many and various sources. As the saying goes, we are often able to move science forward because we are "standing on the shoulders of giants." In my case, I have been extremely fortunate to cross paths with many great scientific minds whose collective influence on my thinking cannot be overstated.

First, I am greatly indebted to my scientific mentors. My first direct experience in ecological science was as a field assistant working in an old-growth forest on Mont Saint-Hilaire, Québec, as part of a project led by Martin Lechowicz, Marcia Waterway, and Graham Bell of McGill University. I was part of a team of mostly inexperienced undergraduate students ripping apart sedges in the lab to make genetic clones and then planting these clones up and down kilometer-long transects in the forest. We eagerly awaited visits to the field site from our mentors—Marcia taught us how to identify everything, Graham laid out the theoretical context, and Marty covered everything in between. Graham's view of the forest plants as being conceptually not so different from his test tubes of evolving algae in the lab at first struck most of us as oversimplified to the point of eliminating all the beauty and mystery from the natural world. But the appeal of focusing on general processes where others saw a near-insurmountable web of details stuck with me, as did Marty and Marcia's insistence that our theoretical models be molded to the reality of the natural world, and not vice versa. The conceptual framework of this book is probably not so different from the one I was first exposed to while on the "*Carex* crew" in the 1990s.

My PhD advisors, Peter Marks and Monica Geber, fostered deep reflection on both empirical and theoretical perspectives in science and, perhaps more important, provided the freedom and unwavering support to intellectually wander the conceptual terrain of ecology and evolution. Stephen Ellner sparked my interest in theory, and the clarity of his thinking remains an inspiration for me today. One more person deserves mention here, although he might be surprised to read this. During our first year as PhD students, Sean Mullen and I were in the introductory biology teaching assistants' room waiting to see if any students would come see us, and he made an offhand comment along the lines of "it would be super cool to know whether patterns of genetic variation in your forest plants correspond to the community-level patterns of interest." That one remark helped set the direction for much of my research for the next 10 years, in particular my interest in integrating community ecology and population genetics, which underpins the conceptual framework in this book. Thanks, Sean.

Another major source of stimulus came from Stephen Hubbell's (2001) book on neutral theory, which essentially represents the importation into community ecology of one specific set of models from population genetics—those excluding selection. In some ways, the present book represents the next logical step: the importation and addition of models involving selection. Janis Antonovics, Bob Holt, and Joan Roughgarden have also made the basic argument that many ecological processes are closely analogous to evolutionary processes, and personal meetings with each of these giants of ecology and evolution provided inspiration at various points along the way.

I have probably received the most encouragement to turn my ideas into a book from students of community ecology or fellow scientists, most of whom were "junior," circa 2010 or before. For providing a testing ground, critical feedback, and substantial encouragement, I have several groups of people to thank: the students and fellow instructors in three courses in community ecology (two at the University of British Columbia, one at Université de Sherbrooke); the students in my lab group, in Margie Mayfield's lab group (University of Queensland), and in various discussion groups; and the many students I have met during seminar visits in too many places to name. These students and their successors are one of my primary target audiences.

A fellow postdoc at the National Center for Ecological Analysis and Synthesis, John Orrock, was coauthor and cobrainstormer on a book chapter that presented an initial sketch of the ideas in this book (Vellend and Orrock 2009), and Anurag Agrawal was very supportive in inviting submission of a more developed paper to *Quarterly Review of Biology* (Vellend 2010). Prior or subsequent to publication of the *QRB* paper, I received especially encouraging and/or constructively critical comments (even if many were quite brief and not necessarily positive) from Peter Adler, Bea Beisner, Marc Cadotte, Jérôme Chave, Jon Chase, Jeremy Fox, Amy Freestone, Jason Fridley, Tad Fukami, Nick Gotelli, Kyle Harms, Marc Johnson, Jonathan Levine, Chris Lortie, Brian McGill, Jason McLachlan, Christine Parent, Bob Ricklefs, Brian Starzomski, James Stegen, and Diego Vázquez. No doubt I am missing some important people from this list, for which I apologize. The University of Queensland, and my sabbatical host there, Margie Mayfield, provided an inspiring setting where the bulk of the writing of this book was completed.

Finally, during the writing of the book I received invaluable input from many people in the form of data, advice, or critical feedback. Raw data for analyses and/or graphics were generously provided by Véronique Boucher-Lalonde, Will Cornwell, Janneke HilleRisLambers and Jonathan Levine, Carmen Montaña, Laura Prugh, Adam Siepielski, Josie Simonis, Janne Soininen, and Caroline Tucker and Tad Fukami. Feedback on particular issues or sections of the book was provided by Jeremy Fox, Monica Geber, Dominique Gravel, Luke

Harmon, Liz Kleynhans, Nathan Kraft, Geoffrey Legault, Jonathan Levine, and Andrew MacDonald. Andrew MacDonald is also responsible for the presentation of R code in a much nicer format than I was capable of producing myself and for helping my programming skills inch slightly toward the twenty-first century (I still have a long way to go). Finally, I owe a very special thanks to several people who read and provided excellent feedback on the book in its entirety: Véronique Boucher-Lalonde, Bob Holt, Marcel Holyoak, Geneviève Lajoie, Andrew Letten, Jenny McCune, Brian McGill, and Caroline Tucker. The book is far better than it would have been without the input from all these people, and I am profoundly grateful for the generosity of everyone who has helped out in some way.

The Theory of
Ecological Communities

CHAPTER 1

Introduction

Many budding ecologists have their imaginations captured by a seemingly simple question: why do we find different types and numbers of species in different places? The question is the same whether the setting is birds in the forest, plants along a mountainside, fish in lakes, invertebrates on a rocky shore, or microbes in the human body. Some parts of the answer to this question are glaringly obvious just from a short walk more or less anywhere on earth. Strolling through any city or town in eastern North America, we can see that the plant species growing in sidewalk cracks and dry roadsides are different from those growing in wet ditches, which are different still from those growing in wooded parks. Some birds reach very high abundance in dense urban areas, while others are found exclusively in wetlands or forests. So, we can observe everyday evidence that environmental variation selects for different species in different places (Fig. 1.1).

As we begin to look more closely, however, the story is not so simple. Some places that seem to present near-identical environmental conditions are nonetheless home to very different sets of species. Some pairs of species seem to live in very similar types of environments but almost never in the same physical place. Two places experiencing a very similar disturbance event (e.g., a drought or fire) subsequently follow very different successional trajectories. A hectare of one type of forest might contain 100-fold more species than a hectare of another type of forest. A major scientific challenge is thus to devise theories that can explain and predict such phenomena. Over the past 150 years ecologists have risen to this challenge, devising hundreds of conceptual or theoretical models that do just this. However, because almost every such model

Figure 1.1. The east-facing slope of Mont Saint-Joseph in Parc national du Mont Mégantic, Québec, illustrating spatial relationships between environmental conditions and community composition. The cold, upper part of the slope (~850–1100 m above sea level) is boreal forest (dark coloration) dominated by balsam fir (*Abies balsamea*). The lower slope is deciduous forest (light coloration) dominated by sugar maple (*Acer saccharum*). The photo was taken in springtime (8 May 2013), prior to the flushing of deciduous leaves. The foreground is relatively flat terrain (~400 m a.s.l.) composed mostly of a patchwork of young forest stands on private land, with a variety of different tree species. From left to right, the image spans roughly 4 km.

is relevant to at least one type of community somewhere on earth, the list of explanations for community patterns gets only ever longer, never shorter.

We are thus faced with a serious pedagogical challenge: how to conceptually organize theoretical ideas in community ecology as simply as possible to facilitate ecological understanding. We have for a long time organized ecological knowledge (in textbooks or other synthetic treatments) according to subareas into which researchers have self-organized rather than fundamental ecological processes that cut across these subareas. For example, a treatment of plant community ecology might have sections on herbivory, competition, disturbance, stress tolerance, dispersal, life-history tradeoffs, and so on (Crawley 1997, Gurevitch et al. 2006). Similarly, a conceptual treatment of community ecology might present many competing theories: island biogeography, priority effects, colonization-competition models, local resource–competition theory, neutral theory, metacommunity theory, and so on (Holyoak et al. 2005, Verhoef and Morin 2010, Morin 2011, Scheiner and Willig 2011, Mittelbach 2012). As a result, if each student in an undergraduate or graduate class is asked to write down a list of processes that can influence community composition and diversity (I have done this several times), the result will be a long list from each student, and collectively no fewer than 20–30 items.

The central argument to be developed in this book is as follows. Underlying all models of community dynamics are just four fundamental, or "high-level," processes: selection (among individuals of different species), ecological drift, dispersal, and speciation (Vellend 2010). These processes parallel the "big four" in evolutionary biology—selection, drift, migration, and mutation—and they allow us to organize knowledge in community ecology in a simpler way than by using the conventional approach. What seems like a jumble of independent theoretical perspectives can be understood as different mixtures of a few basic ingredients. By articulating a series of hypotheses and predictions based on the action of these four processes, we can thus build a general theory of ecological communities. As explained further in Chapter 2, the theory does not apply equally to all topics under the broad umbrella of community ecology. For example, models of species on the same trophic level interacting via competition and/or facilitation (sometimes called "horizontal" communities) fall cleanly within the theory, whereas models involving trophic interactions fit within the theory largely to the extent that they make predictions concerning properties of horizontal components of the larger food web (which they often do). Nonetheless, following the tradition set by MacArthur and Wilson (1967, *The Theory of Island Biogeography*) and Hubbell (2001, *The Unified Neutral Theory of Biodiversity and Biogeography*), I call my theory and therefore my book *The Theory of Ecological Communities*.

1.1. WHAT THIS BOOK IS

My overarching objective in this book is to present a synthetic perspective on community ecology that can help researchers and students better understand the linkages among the many theoretical ideas in the field. The initial sketch of these ideas was presented in Vellend (2010), and this book is a fully fleshed-out version of the theory, reiterating the key points of the earlier paper but going well beyond it in many ways:

- First, I more thoroughly place the theory of ecological communities in historical context (Chap. 3), and I present a novel perspective (gleaned from philosopher Elliott Sober) on why high-level processes (in this case selection, drift, dispersal, and speciation) represent an especially appropriate place to seek generality in community ecology (Chap. 4).
- I describe in detail how a vast number of different hypotheses and models in community ecology fit as constituents of the more general theory (Chap. 5).
- I provide simple computer code in the R language that (i) generates predictions for empirical testing, (ii) illustrates how changing a few basic "rules" of community dynamics reproduces a wide range of well-known models, and (iii) allows readers to explore such dynamics on their own (Chap. 6).
- After outlining some key motivations and challenges involved in empirical studies in ecology (Chap. 7), I then put the theory of ecological communities to work by systematically articulating hypotheses and predictions based on the action of selection (Chap. 8), drift and dispersal (Chap. 9), and speciation (Chap. 10), in each case evaluating empirical evidence supporting (or not) the predictions. In essence, Chapters 8–10 serve to reframe the corpus of empirical studies in community ecology according to a general theory that is considerably simpler than typically found in a textbook treatment of the discipline.

- Chapters 11 and 12 present some overarching conclusions and a look to the future.

1.1.1. Reading This Book as a Beginner, an Expert, or Something in Between

This book is aimed at senior undergraduate students, graduate students, and established researchers in ecology and evolutionary biology. It is the book I would have liked to read during grad school. I believe it presents the core conceptual material of community ecology in a new and unique way that makes it easier to grasp the nature of the key processes underlying community dynamics and how different approaches fit together. This has been my experience in using it as a teaching tool. I also hope to stimulate established researchers to think about what they do from a different perspective, and perhaps to influence how they teach community ecology themselves. Thus, I approached the writing of the book with the dual goals of pedagogy (beginning-student audience) and advancing a new way of thinking about theory in community ecology (expert audience). I suspect that readers who are somewhere on the pathway from beginner to expert—that is, grad students—have the most to gain from reading this book.

A pervasive challenge in scientific communication (including teaching) is to keep the most knowledgeable members of an audience engaged without "losing" those with the least preexisting knowledge of the topic. Readers can get the most out of this book if they are already somewhat familiar with the kinds of community-level patterns of species diversity and composition that ecologists aim to explain, as well as some of the factors commonly invoked to explain such patterns—environmental conditions, competition, disturbance, and so on. I begin explanations at a fairly basic level and provide what I consider the essential background (Chaps. 2–3), but even so, a full understanding of various historical advances in ecology (Chap. 3) and some of the more sophisticated empirical studies (Chaps. 8–11) requires delving into the primary literature. At the other end of the spectrum, expert readers will no doubt encounter sections they can skim, but I hope that all chapters of the book contain enough novel perspectives, approaches, or modes of traversing well-trodden ground to engage even the most expert reader. If you are an expert and pressed for time, you may choose to skip to the end of Chapter 3 (Sec. 3.4), where I begin the transition from background material to the details of my own distinct perspective and theory. Feedback on earlier versions of the book suggested that experts will find the most "new stuff" in the latter part of the book (Chaps. 8–12).

1.1.2. Unavoidable Trade-Offs

This book covers a very broad range of topics (models, questions, methods, etc.), which necessarily involves a trade-off with detail in several respects. First, the depth to which I explore each individual topic is limited. So, while readers will learn, for example, about the strengths and weaknesses of different approaches to testing for signatures of ecological drift or spatially variable selection, they will not learn all the detailed ins and outs of how to implement particular empirical methods. I am not myself an expert on all such details, and even for topics I do know quite well, I have deliberately limited the detail so as not to distract from the big-picture conceptual issues on which I want to focus. Plenty of references are provided for readers interested in digging deeper. Second, I present very few formal statistics, despite their ubiquity in ecological publications. I report a great many empirical results from the literature, but almost entirely in graphical form, allowing readers to see for themselves the patterns in the data. Interested readers can consult the original publications for p-values, slopes, r^2, AIC, and the like. Finally, I cannot claim to have cited the original paper(s) on all topics. My emphasis has been on communicating the ideas rather than tracing each of their histories to the origin, although I do dedicate a whole chapter to the history of ideas, and I hope I have managed to give credit to most of those papers considered "classics" by community ecologists.

1.1.3. Sources of Inspiration

By way of ensuring that I have appropriately credited the ideas that form the basic premise of this book, I end this introductory chapter by acknowledging those publications that inspired me by calling attention to the striking conceptual parallels between population genetics and community ecology (Antonovics 1976, Amarasekare 2000, Antonovics 2003, Holt 2005, Hu et al. 2006, Roughgarden 2009). Many additional researchers have taken notice of these parallels, especially following the importation into ecology of neutral theory from population genetics (Hubbell 2001). That said, I can say from experience that most community ecologists have *not* thought of things in this way, and there has been no systematic effort to find out whether it's possible to reframe the bewildering number of theories, models, and ideas in community ecology as constituents of a more general theory involving only four high-level processes. This book is my attempt to do so.

PART I
APPROACHES, IDEAS, AND THEORIES
IN COMMUNITY ECOLOGY

How Ecologists Study Communities

The next three chapters serve three main purposes: (i) to establish the domain of application of the theory of ecological communities, (ii) to describe the basic community patterns of interest, and (iii) to place the book in historical context. The present chapter is aimed largely at goals (i) and (ii), but as a by-product it also begins to address goal (iii). Historical context is addressed more fully in Chapters 3 and 4.

Ecologists study communities in a variety of ways. In the very same study system (e.g., temperate lakes), one ecologist might focus on the phytoplankton community, while another focuses on the interaction between zooplankton and a dominant fish species. One study might focus on the processes that determine community structure in a single lake, while another describes patterns across several lakes in one landscape, or in thousands of lakes across an entire continent. Finally, one researcher might be primarily interested in understanding why lakes vary according to *how many* species they contain, while another is more interested in why lakes vary according to *which* species they contain. Thus, any study in community ecology must establish from the outset at least three things: the focal set of species, the spatial scale of analysis, and the community properties of interest. The following two sections establish the domain of application of the theory of ecological communities according to the delineation of a focal set of species (Sec. 2.1) and spatial scales of interest (Sec. 2.2). Relative to the traditional view of community ecology focused on interactions between species at a local scale (Morin 2011), the domain of application here is in one sense narrower (focusing largely on single trophic levels) and in another sense broader (focusing on all scales of space and time). Having established the domain of application, I then describe the properties of communities that ecologists strive to understand (Sec. 2.3).

2.1. DIFFERENT WAYS OF DELINEATING ECOLOGICAL COMMUNITIES

All scientific endeavors must define their objects of investigation, and so community ecologists must define their ecological communities. As a theoretical ideal, it is useful to consider the complete set of organisms belonging to all species (viruses, microbes, plants, animals) living in a particular place and time as an ecological community *sensu lato* (Fig. 2.1a). In practice, however, this theoretical ideal is almost never met. Researchers more or less always begin their studies by focusing on a subset of the full community, chosen on the basis of taxonomy, trophic position, or particular interactions of interest (Morin 2011). Recognizing the many ways researchers delimit communities, we can use the maximally inclusive definition of a community as "a group of organisms representing multiple species living in a specified place and time" (Vellend 2010; see also Levins and Lewontin 1980). Once a researcher has chosen a group of organisms as the focal community, all other components of the ecosystem— biotic and abiotic— are then conceptually externalized, in the sense that they may be ignored completely, or incorporated into an investigation as variables that may influence the object of study, without being formally a part of the object of study itself (Fig. 2.1).

Focal groups of species to be included in a community of interest can be defined in many ways. Some of the earliest studies in community ecology treated "plant communities" (Clements 1916) and "animal communities" (Elton 1927) as separate, if interacting, objects of study. In contemporary ecology, studies of "food webs" (McCann 2011) focus on feeding relationships, often ignoring differences among species within trophic groups, and externalizing nonfeeding interactions and even some feeding interactions (e.g., nectar consumption by pollinating insects) (Fig. 2.1b). Studies of "mutualistic networks" (Bascompte and Jordano 2013) focus on two sets of interacting species, such as plants and their pollinators or mycorrhizae, externalizing everything else (Fig. 2.1c). Studies also often focus on a small number of strongly interacting species— "community modules" *sensu* Holt (1997)—such as particular consumer-resource pairs (e.g., the lynx and the hare) (Fig. 2.1d).

Finally, one can choose to focus on species at a particular trophic level (e.g., plants) or in a particular taxon (e.g., birds or insects), again externalizing everything else (Fig. 2.1e). Ecologists have referred to such a unit of study (or something like it) as an "assemblage" (Fauth et al. 1996), a "guild" (Root 1967), a set of "species having similar ecology" (Chesson 2000b), or a "horizontal community" (Loreau 2010). These terms are all decidedly lacking in the pizazz and the admirable self-defining quality of the other terms in Figure 2.1, so throughout this book, for lack of a better term, I will simply refer to them as *ecological* communities and, occasionally, *horizontal* communities when the distinction is helpful.

2.1.1. A Focus on Horizontal Ecological Communities

The theory of ecological communities applies unambiguously to horizontal ecological communities, and so this book is largely about horizontal communities, which are also the focus of a sizeable proportion of studies in ecology as a whole (see Chap. 7). I am a plant ecologist, and my empirical studies in community ecology focus on the set of plant species found in different places and at different times (e.g., Vellend 2004; Vellend et al. 2006, 2007, 2013). This is the bias I bring to the table, and not surprisingly, the theory I will describe applies nicely to plant communities. However, it applies equally to any set of species sharing common needs in terms of resources or space, such as phytoplankton, sessile intertidal invertebrates, seed eaters, decomposers, predatory insects, or songbirds. Importantly, species in such communities interact not only via competition, which has traditionally received the most attention from ecologists, but also via facilitation (positive interactions) and any number of positive or negative indirect interactions via other biotic or abiotic components of the ecosystem (Holt 1977, Ricklefs and Miller 1999, Krebs 2009). The theory here is not just about competition.

Within horizontal communities, individuals of different species share similar abiotic and biotic constraints on fitness, and community dynamics are closely analogous to the evolutionary dynamics of genotypes in a single species' population (Nowak 2006; see also Chap. 5). Fitness can be quantified in a comparable way across individuals of the same or different species, and so many theoretical models from population genetics apply just as easily to species in communities as they do to alleles or genotypes in populations (Molofsky et al. 1999, Amarasekare 2000, Norberg et al. 2001, Vellend 2010). Such models are based on just four high-level processes: selection, drift, mutation, and gene flow in population genetics; selection, drift, speciation, and dispersal in community ecology. For objects of study that include multiple trophic levels, one can still identify the same four high-level processes, but the analogy is weaker.

Synthetic treatments have already appeared in this very monograph series on food webs (McCann 2011), mutualistic networks (Bascompte and Jordano 2013), and consumer-resource interactions (Murdoch et al. 2013). The present book completes the picture in Figure 2.1 by providing a conceptual synthesis of the literature on horizontal ecological communities. I build on previous monographs that have covered particular models and theories in horizontal community ecology (MacArthur and Wilson 1967, Tilman 1982, Hubbell 2001). Whether and how the conceptual frameworks for food webs, mutualistic networks, interaction modules, and horizontal communities can be truly merged—and not simply squeezed side by side into the

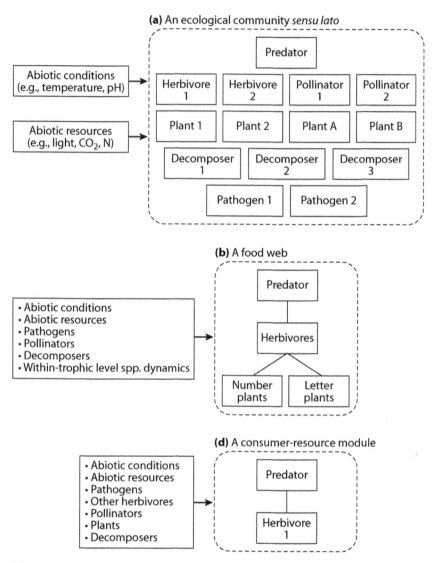

Figure 2.1. Different ways of defining objects of study in community ecology for a hypothetical terrestrial ecosystem. Each diagram represents the same system, but with the direct object of study (whatever is inside the dashed-line box) defined in different ways. Solid lines indicate interactions between species under consideration (omitted from (a) for simplicity); solid-line boxes outside of the dashed-line box indicate all components of the ecosystem that are externalized. Plant species 1 and 2 ("number plants") belong to a different functional group than species A and B ("letter plants"); species within such functional groups (e.g., herbs vs. shrubs) are often lumped in food-web analyses (b).

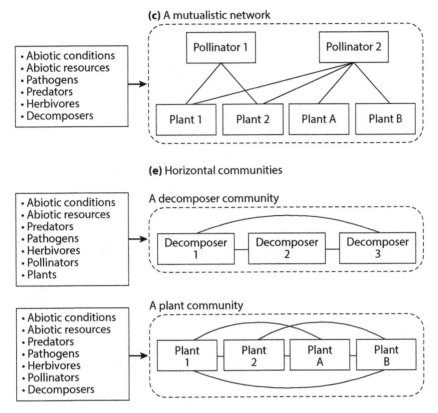

Figure 2.1. *Continued*

same package or combined in some specific contexts (there are lots of examples of both)—remains to be seen. Before proceeding, I do wish to emphasize that this focus does not marginalize the importance of trophic interactions or any other process or variable treated as external to the object of study. Rather, as indicated previously, it simply treats consumers or pathogens or mutualists as being outside the direct study object of interest—a component of the biotic environment that potentially exerts strong selection on the focal community and that might itself respond to changes in this focal community (Fig. 2.1).

2.2. THE UBIQUITOUS ISSUE OF SCALE

In addition to studying very different sets of species, researchers also study communities at many different spatial scales. Some definitions of an ecological community (reviewed in Morin 2011) include species interactions as a requirement, thus placing an upper limit on the spatial extent of a community. I see no objective way to define such a spatial limit, so I prefer to leave species interactions and any spatial restriction out of the definition of an ecological community. Following Elton (1927), I consider the concept of a community "a very elastic one, since we can use it to describe on the one hand the fauna of equatorial forest, and on the other hand the fauna of a mouse's cæcum." As such, the theory of ecological communities applies to analyses of community properties (see Sec. 2.3) at any scale of space or time, which includes studies not typically described under the banner of community ecology but, rather, as biogeography, macroecology, or paleoecology.

Even with a spatially elastic definition of communities, one must remain cognizant that processes and patterns observed at one scale may be quite different from those observed at another scale, and that a community may be influenced by processes operating at many scales (Levin 1992). A tree in a forest may experience competition only from its neighbors within a few meters, it may be pollinated by insects traveling hundreds of meters, and its growth might be affected by climatic fluctuations originating in altered water circulation in the southern Pacific Ocean thousands of kilometers away. As will become clear in subsequent chapters, some processes—e.g., negative frequency-dependent selection—can result from highly localized species interactions or from trade-offs involving dispersal at larger scales. The key point here is that it is rarely if ever possible to define one "correct" scale for studying ecological phenomena of interest (Levin 1992), especially if we are interested in multiple interacting processes that might determine community structure and dynamics.

Although the focal scale of observation can clearly vary continuously from tiny study plots to entire continents, for sake of convenience community ecologists often recognize discrete scales, such as "local" (the smallest), "global" (the largest), and "regional" (somewhere in between) (Ricklefs and Schluter 1993b, Leibold et al. 2004). For many studies, there is no need for scaling terminology at all, in that areas of a narrow size range are being compared. These areas might be highly variable in size from study to study—1- m² plots, islands of several square kilometers, or portions of a continent of hundreds of square kilometers—but there is still no need to assign them a label. In other cases, researchers refer to patterns or processes at multiple scales, most often those occurring within the smallest focal area and those operating at larger scales. In these cases, scaling terms are quite handy. Following convention in the literature, I refer to these as "local" and "regional," respectively, while recognizing that these terms do not have precise meanings beyond the fact that one is nested within the other (see also Chap. 5). An ensemble of local communities is a "metacommunity."

2.3. THE PROPERTIES OF ECOLOGICAL COMMUNITIES THAT WE WANT TO EXPLAIN AND PREDICT

For any focal community and spatial scale(s), ecologists have defined a variety of different community-level properties of interest. Most generally, we are interested in the number of species (species richness), the equitability of abundances and the variety of traits among species (species/trait diversity), the identities and average trait values of species (species/trait composition), and the relationships of these properties to site characteristics. This section describes the quantification of these community properties, which we will refer to frequently in subsequent chapters.

The basic quantitative description of a community is a vector of species abundances, which we can call $\mathbf{A}$. For a community of four species, with abundances of $a_1 = 4$, $a_2 = 300$, $a_3 = 56$, and $a_4 = 23$, $\mathbf{A} = [4, 300, 56, 23]$. The abundances might be the number of adult trees of maple, beech, ash, and pine, or of four species of woodpecker in a given site. This is equivalent to thinking of the abundance of each of S species as a "state variable," with the state of the community being its position in the resulting S-dimensional space (Lewontin 1974). Most observational studies in community ecology include data from multiple sites or local communities, in which case the raw data are represented by a species × site matrix whose elements are species abundances (Fig. 2.2), many of which are likely to be zeros. This matrix represents the "metacommunity" and is made up of concatenated species abundance vectors, one for each of j sites ($\mathbf{A}_1$, $\mathbf{A}_2$ . . . $\mathbf{A}_j$). Armed with such data (and no additional data), we can calculate what we might consider "first-order" community properties, as follows.

First-Order Properties of Single Communities (Fig. 2.2)

Species richness: The number of species in the site (i.e., the number of nonzero elements in $\mathbf{A}$). When comparing sites in which different numbers of individuals were observed, researchers often standardize species richness across sites by calculating the number of species observed in repeated random samples of a given number of individuals in a given site, a procedure called "rarefaction." Unstandardized species richness is sometimes called "species density" (Gotelli and Colwell 2001).

Species evenness/diversity: Any index calculated from the vector of species abundances $\mathbf{A}$ in which, all else being equal, a more even distribution of abundances across species leads to higher values. For example, a forest in which total tree abundance is split 50:50 between two species has higher evenness than a forest in which one of the two species abundances greatly exceeds the other. Common indices are the Shannon-Weiner Index, Simpson's Index, and various kinds of entropy calculations (Magurran and

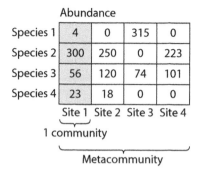

Abundance

	Site 1	Site 2	Site 3	Site 4
Species 1	4	0	315	0
Species 2	300	250	0	223
Species 3	56	120	74	101
Species 4	23	18	0	0

1 community

Metacommunity

For site 1:

Species richness = 4

Simpson's evenness =
$1/\Sigma \text{freq}_i^2 = 1/\Sigma((4/383)^2+(300/383)^2+(56/383)^2+(23/383)^2) = 1.57$

Species abundance distribution
(shown as a rank-abundance plot):

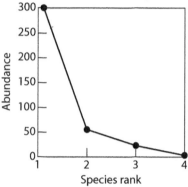

Figure 2.2. The basic quantitative description of a metacommunity. The vector of species' abundances, $\mathbf{A}_1$ (gray-shaded rectangle), represents the multivariate "composition" of the community in site 1, while the matrix represents the metacommunity of four sites. Calculations of other first-order properties of the community at site 1 are shown to the right. *freq*$_i$ is the frequency of species *i* (its abundance divided by the sum of species' abundances for that site).

McGill 2010). These indices are typically based on the frequencies, freq$_i$ = $a_i/\Sigma(a_i)$, of each species *i*.

Species composition: The vector of abundances (sometimes just recorded as presence/absence) itself, unmodified (**A**), can be considered a multivariate property of the community that one might wish to explain or predict.

Species abundance distribution: Regardless of which species has which abundance, the distribution of these abundances (e.g., is it log-normal or some other shape?) has been of great interest as a community property (McGill et al. 2007).

First-Order Properties of Multiple Communities (i.e., a Metacommunity)

Beta diversity: Species composition is often studied by first calculating indices that represent the degree of dissimilarity in species composition, or beta diversity, between sites. To quantify beta diversity, a single number can be calculated for the entire set of sites, or more commonly, an index is calculated for each pairwise combination of sites. There is a very long list

Abundance

	Site 1	Site 2	Site 3	Site 4		Trait 1	Trait 2	Trait 3	Trait 4
Species 1	4	0	315	0		0.2	320	0.5	20
Species 2	300	250	0	223		0.6	298	0.1	16
Species 3	56	120	74	101		0.9	412	0.1	26
Species 4	23	18	0	0		1.3	300	0.2	21

	Site 1	Site 2	Site 3	Site 4
Site characteristic 1	10	1	7	16
Site characteristic 2	0.01	0.4	0.2	0.5
Site characteristic 3	90	92	95	97
Site characteristic 4	12	0.1	0	5

Figure 2.3. The three data tables needed to calculate various second-order community properties, either incorporating species characteristics (traits) or site characteristics (e.g., environmental variables). Traits are assumed to be fixed at the species level (i.e., not variable within species among sites).

of possible indices (Anderson et al. 2011). Without getting into quantitative details, pairs of sites sharing similar abundance vectors (e.g., sites 1 and 2 in Fig. 2.2) show low beta diversity, while pairs with very different abundance vectors (e.g., sites 1 and 3) show high beta diversity. For example, beta diversity between a tropical and a temperate forest will be much higher than beta diversity between two nearby plots in the temperate forest. This difference can be due both to turnover in the identities of species in each site, as well as to differences in abundances of species present in both sites.

Characterizations of pattern in ecological communities often draw on two additional kinds of data. First, we can incorporate the characteristics ("traits") of each species into calculations of community properties for either single or multiple communities (Weiher 2010; Fig. 2.3). For example, rather than asking about how many species (species richness) and which species (species composition) live in different places, we can ask how the variance in body size among species (trait diversity) or the average body size across species (trait composition) varies from place to place (McGill et al. 2006). Phylogenetic relationships among species represent a special case of incorporating species-level characteristics, with which we can calculate the degree of relatedness among species as an index of "phylogenetic diversity" (Vellend et al. 2010).

Second, we can use site characteristics (e.g., their area, environment, or geographic isolation) to quantify relationships with first-order properties, and such relationships then represent patterns in their own right requiring explanation.

For example, we might calculate parameters of the statistical relationship between species richness and area or productivity and then seek to understand how those parameters vary from case to case (Rosenzweig 1995). Although not all site characteristics (e.g., temperature, geographic isolation, presence of a predator) fall readily under the heading "environmental variable," for simplicity I will often use the term "environment" to refer them as a group. Given the incorporation of data that go beyond the basic species × site matrix, we can think of the patterns just described as "second-order" community properties.

Second-Order Community Properties Incorporating Species Characteristics

Trait diversity: Several indices quantify the degree of within-community variation among species in trait values, for either single or multiple traits. Traits are often described as "functional" in the sense that they have some impact on fitness (Violle et al. 2007), so such indices are frequently described as reflecting "functional diversity" (Laliberté and Legendre 2010, Weiher 2010).

Trait composition: The community-level average value of a given trait is essentially one way to quantify community composition (Shipley 2010).

Second-Order Community Properties Incorporating Site Characteristics

Diversity-environment relationships: The relationship between local diversity (any first- or second-order metric) and a particular site characteristic. Frequently studied site characteristics include the area over which a survey was done (species-area relationships) and any number of "environmental" variables such as productivity, disturbance history, elevation, latitude, pH, geographic isolation, or soil moisture availability (Rosenzweig 1995).

Composition-environment relationships: The strength and nature of relationships between species composition (including trait or phylogenetic composition) and any site characteristics of interest. Such analyses can be implemented in many different ways (Legendre and Legendre 2012), including analyses that aim to predict pairwise site-to-site indices of beta diversity using pairwise differences among sites in certain characteristics (e.g., the geographic distance between sites). These are discussed further in Chapter 8.

It is important to emphasize that there is an extremely large number of ways of calculating each type of community property, and an even larger number of ways of analyzing those properties. These have been described in detail elsewhere (Magurran and McGill 2010, Anderson et al. 2011, Legendre and Legendre 2012). I have presented here the basic set of conceptually distinct patterns one might quantify in communities.

To summarize, the theory of ecological communities applies most clearly to horizontal ecological communities, and it applies to any scale of space or time. Community ecologists describe patterns in a variety of first- and second-order community properties, and in subsequent chapters I explore how all these properties can be understood as the outcome of four high-level processes: selection, drift, speciation, and dispersal.

CHAPTER 3

A Brief History of Ideas in Community Ecology

Ideas do not arise in a vacuum. This book was inspired both by a perceived "mess" of loosely related models and patterns in community ecology (McIntosh 1980, Lawton 1999), as well as some conceptual developments in ecology and evolutionary biology that pointed the way to what I consider a very general theory that can help contain the mess (Mayr 1982, Ricklefs 1987, Hubbell 2001, Leibold et al. 2004). In addition to my broad goal in this chapter of putting the theory of ecological communities in historical context, my aim is to communicate both how community ecology came to be perceived as a mess and where the building blocks of my own theory originated. I do so by providing a brief history of the research traditions in ecology most relevant to horizontal communities. Along the way, if you start to feel confused about how all the historical pieces of community ecology fit together, that is indeed one of the points I wish to make, and is a problem the rest of this book aims to solve. Experts familiar with the history of community ecology may wish to skip to Sections 3.4 and 3.5, which present a synthetic, forward-looking perspective.

The history of community ecology does not involve a linear sequence of events. For any current research area (e.g., metacommunities or trait-based community analyses), one can identify numerous intellectual strands extending backward in time to different origins. Likewise, most foundational ideas (e.g., the competitive exclusion principle or the individualistic concept) have had an influence on many different current topics of research (McIntosh 1985, Worster 1994, Kingsland 1995, Cooper 2003). Therefore, any one person's historical account will differ from that of others. In addition, because the basic subject matter of ecology involves commonplace phenomena such as the distributions and behaviors of plants and animals, core ideas in ecology can be traced back thousands of years (Egerton 2012). Many nineteenth-century scientists and natural historians, such as Alexander von Humboldt (1769–1859), Charles Darwin (1809–1882), and Eugenius Warming (1841–1924), could easily be considered community ecologists (among other things) by modern standards. Nonetheless, to understand how the different pieces of present-day community ecology fit together, we can stick largely with more recent conceptual developments.

The brief historical sketches I present here extend back no further than a century or so. I have not aimed to provide a comprehensive historical account nor to mention all important contributions, as several excellent historical treatments of ecology collectively do this (McIntosh 1985, Worster 1994, Kingsland 1995, Cooper 2003, Egerton 2012). With an eye to covering the conceptual ground necessary to understanding modern-day "horizontal" community ecology specifically (see Fig. 2.1), I focus on the development of three themes: (i) making sense of community patterns (Sec. 3.1), (ii) generating and testing predictions from simplified mathematical models (Sec. 3.2), and (iii) examining the importance of large-scale processes (Sec. 3.3). Section 3.4 focuses on a series of debates and waves of interest in various topics in community ecology over the past 50 or so years, from which the building blocks of the theory of ecological communities emerged. I focus this chapter largely on conceptual approaches and developments, with minimal empirical content. Empirical studies are the subject of Chapters 7–10.

3.1. MAKING SENSE OF COMMUNITY PATTERNS OBSERVED IN THE FIELD

For well over a century, field biologists have been characterizing patterns in ecological communities and trying to draw theoretical inferences from the resulting data. One of the earliest theoretical debates in ecology concerned the question of whether communities in nature could be recognized as discrete entities. Frederic Clements (1916), an American plant ecologist, said yes. Clements held that a community was an integrated entity within which species were as interdependent as organs in a human body. According to this point of view, change in species composition along an environmental gradient is not gradual but, rather, abrupt (Fig. 3.1a). Because of this strong interdependence among species within a community, moving up a mountainside of mature forest, for example, one could find oneself in community type 1 or 2 but rarely if ever in a transitional community type (Fig. 3.1a).

Clements's ideas aligned nicely with the tradition of vegetation classification, which was a major focus of botanists' efforts in Europe in the early twentieth century, as typified by the "Zurich-Montpellier" approach pioneered by Josias Braun-Blanquet and colleagues (Braun-Blanquet 1932). The basic data

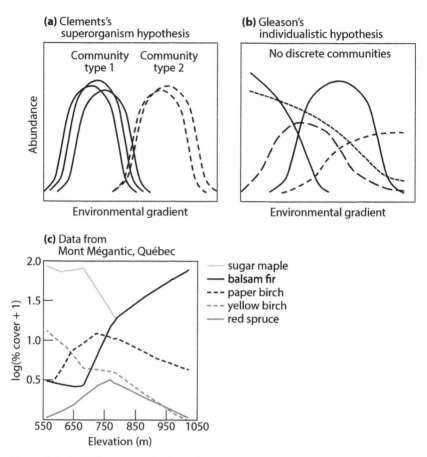

Figure 3.1. (a, b) Two competing hypotheses regarding species distributions along environmental gradients, and thus the organization of species into communities (i.e., particular points along the *x*-axis). (c) Locally weighted scatter plot smoothing (LOESS) curves (tension = 0.7) for the five most abundant tree species in 48 vegetation plots along an elevational gradient at Mont Mégantic, Québec (data from Marcotte and Grandtner 1974). These data illustrate gradual change in community composition along the gradient, thus supporting hypothesis (b).

involved plant community surveys, with subsequent efforts aimed at organizing study plots into a hierarchical vegetation classification scheme (each plot was assigned to a particular vegetation "type"), thus implicitly assuming that ecological communities are discrete entities.

Opposition to Clements's view of an ecological community as a "superorganism" is most often associated with Henry Gleason, who argued that each species responded in a unique way to environmental conditions (Fig. 3.1b).

According to this point of view, the set of species one finds in a given place results more from species-specific or "individualistic" responses to various environmental factors than from strong interdependence among species (Gleason 1926). Support for this supposition came later from data showing gradual variation in community composition along environmental gradients (e.g., elevation) rather than abrupt transitions from one community type to the next (Whittaker 1956, Curtis 1959; Fig. 3.1c). This reality forced ecologists to operationally define communities as the set of species in sometimes arbitrary units of space, as I have done in this book (see Chap. 2).

Until the 1950s, analyses of community survey data were largely qualitative. Quantitative data were presented in tabular form or in graphs of how species abundances changed along particular gradients (e.g., Fig. 3.1), but conclusions were drawn from qualitative inspection of such tables and graphs (e.g., Whittaker 1956). A clear need for quantitative, multivariate methods of analysis was apparent, and this need was filled by methods developed under the heading of "ordination" (Bray and Curtis 1957). Multivariate ordination aims to place survey plots "in order" based on their multivariate species composition. Such methods begin by considering the abundance of each species as a separate variable measured at each site, such that the "response" of interest is inherently multivariate (i.e., the vector of species abundances described in Chap. 2). Because many species pairs show correlated distribution patterns (positive or negative), ordination methods are typically able to identify and extract a relatively small and manageable number of dimensions along which most variation in community composition occurs (Legendre and Legendre 2012). For example, if we conduct an ordination of only the species × site data used to create Figure 3.1c (i.e., *without* incorporating any information on elevation), the first axis of an ordination analysis would correlate strongly with elevation, given that so many of the species show correlated distribution patterns along this axis. Such methods allow one to ask—quantitatively— which environmental or spatial variables best predict site-to-site variation in community composition (Legendre and Legendre 2012)?

To the extent that different community theories make different predictions about the explanatory power of different variables, the results of multivariate community analyses can in principle allow empirical tests (see Chaps. 8–9). As a relatively recent example, neutral theories (described in Sec. 3.3) predict no direct role of environmental variables (e.g., elevation or pH) in explaining community composition but an important role of spatial proximity among sites. The development and application of new multivariate methods of community analysis has continued unabated for the last 50+ years and characterizes a major thrust of current research (Anderson et al. 2011, Legendre and Legendre 2012, Warton et al. 2015).

As described in Chapter 2, ecologists have documented many other community-level patterns as well, such as species-area relationships, relative abundance distributions, and trait distributions (e.g., body size), and subsequently have sought explanations for such patterns. Many such explanations derive from mathematical models of one sort or another, as described in the next two sections.

3.2. SIMPLIFIED MATHEMATICAL MODELS OF INTERACTING SPECIES

One cannot overstate the massive influence of population modeling in ecology. This is one case in which a major research tradition can be traced back to some singular contributions (Kingsland 1995), in particular the models of interacting species developed independently by Alfred Lotka and Vito Volterra (see also Nicholson and Bailey 1935). Models of this nature can be used to try to make sense of community patterns already observed and to generate new predictions for how community dynamics should proceed under different conditions. The simulation models presented in Chapter 6 fall squarely within this tradition. To understand where these models, and their hundreds of descendants, come from, we must start with simple models of single populations.

Population growth is a multiplicative process. When a single bacterium splits in two, the population has doubled, and when these two cells divide, the population has doubled again, to produce a total of four individuals. If N_t is the population size at time t, and cell division happens in discrete time steps, then $N_1 = N_0 \times 2$, $N_2 = N_1 \times 2 = N_0 \times 2 \times 2$, and so on. For any "reproductive factor" R, $N_{t+1} = N_t \times R$ (Otto and Day 2011). Because the population grows multiplicatively without limit according to this equation (Fig. 3.2a), it is called *exponential population growth*. To make the transition to more complex models smoother, we can define $R = 1 + r$, in which r is the intrinsic rate of population growth. Otto and Day (2011) use the symbol r_d to distinguish this definition of r for discrete-time models from that used in continuous-time models ($r = \log R$), but here I use just r to simplify the notation. If $r > 0$, the population grows, and vice versa. So, $N_{t+1} = N_t(1 + r)$, and

$$N_{t+1} = N_t + N_t r.$$

Of course, populations cannot grow without limit. Although many factors can limit population growth, for a single species the most obvious possibility is the depletion of resources as more and more individuals consume from the same limited supply. In this case, resources should be very abundant when a species is at low density (i.e., there are no organisms to deplete the resource), and so the population can grow exponentially. As the population grows, resources will be depleted, and so population growth should slow. If we define a maximum population size that can be sustained in a given place as K, the "carrying capacity," then population growth should decrease as the population size approaches K. N_t/K expresses how close the population is to K, so $1 - N_t/K$ expresses how *far* the population is from K. We can express the realized population growth as $r(1 - N_t/K)$. If $N_t = K$, the realized population growth is zero, and as N_t approaches zero, the realized population growth approaches r. This scenario is captured by the logistic equation for population growth (Fig. 3.2a):

$$N_{t+1} = N_t + N_t r(1 - N_t/K).$$

The logistic equation represents a minor modification of the exponential-growth equation via the addition of reduced population growth as the population itself gets large and depletes resources. But of course, resources can be depleted (or, in principle, added) by other species. With the addition of a second species, we now need subscripts 1 and 2 to keep track of species-specific variables and parameters (e.g., N_1 and N_2). A simple way to model competition is to express the influence of each individual of species 2 on species 1 as some proportion of the influence species 1 has on itself. We call this parameter the *competition coefficient*, α_{12} (the effect on species 1 of species 2). If an individual of species 2 depletes the resources needed by species 1 at half the rate that species 1 depletes its own resources, $\alpha_{12} = 0.5$. So, if there are N_2 individuals of species 2 in the community, they have the equivalent effect on species 1 as $\alpha_{12} \times N_2 = 0.5 \times N_2$ individuals of species 1. With this assumption, we can now account for resource depletion by competing species in models of the two species population dynamics. Things look more complicated because we have to introduce all the subscripts, but it is really just one small addition to the logistic equation:

$$N_{1(t+1)} = N_{1(t)} + N_{1(t)} r_1 (1 - N_{1(t)}/K_1 - \alpha_{12} N_{2(t)}/K_1);$$
$$N_{2(t+1)} = N_{2(t)} + N_{2(t)} r_2 (1 - N_{2(t)}/K_2 - \alpha_{21} N_{1(t)}/K_2).$$

To model more species, we add an equation for each species and include an additional factor $\alpha_{ij} N_j$ for the effect of each species j on species i.

In Chapter 6 we will explore theoretical dynamics in some models of this nature. For now, suffice it to say that the outcome of competition between species 1 and 2 depends largely on the relative values of the K's and the α_{ij}'s. All else being equal, stable species coexistence is promoted when intraspecific competition is stronger than interspecific competition (i.e., $\alpha_{12} \times \alpha_{21} < 1$) and when the carrying capacities, K_1 and K_2, are not too different (Fig. 3.2b, c). Basic mathematical models of this type for interacting species have been a part of ecology for roughly 100 years, and an enormous number of minor (and perhaps not so minor) modifications have been introduced since then.

3.2.1. The Enduring Influence of Population Modeling in Theoretical and Empirical Community Ecology

A major wave of enthusiasm for mathematical models in ecology swelled in the 1960s and 1970s, largely via the contributions of Robert MacArthur

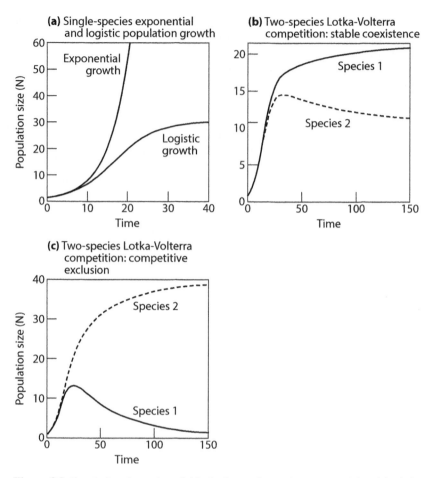

Figure 3.2. Population dynamics of (a) single species under exponential and logistic population growth, and (b, c) two competing species under Lotka-Volterra competition. In all panels, $r = r_1 = r_2 = 0.2$ (see the text for equations). In (b) and (c), $\alpha_{21} = 0.9$, and $\alpha_{12} = 0.8$, such that species 1 has a stronger competitive effect on species 2 than the reverse. For carrying capacities, $K = 30$ for the logistic growth model in (a), for both species in (b), and for species 1 in (c). In (c), $K_2 = 40$, thus giving an advantage to species 2, which overcomes its weaker competitive effect.

and colleagues, including MacArthur's PhD director, G. Evelyn Hutchinson (Kingsland 1995). Many models of competing species include an explicit accounting of the dynamics of resources (e.g., equations representing the dynamics of the limiting nutrients for which plants compete), the results of which help specify the types of trade-offs among species that might promote stable coexistence. For example, if each of two species is (i) limited by a different resource and (ii) takes up the resource by which it is most limited faster than the other species, stable coexistence is possible given certain rates of supply of the two resources (Tilman 1982).

Ultimately, it was realized that regardless of the details of a particular model or natural community, the long-term outcome of competition among species depends on just two key factors (Chesson 2000b). This result can be illustrated by first recognizing that stable species coexistence depends fundamentally on each species having a tendency to increase when its abundance gets very low. Otherwise, we should see competitive exclusion. Even from the original Lotka-Volterra competition model, we can learn that coexistence depends on two key interacting factors: (i) intraspecific competition must be stronger than interspecific competition ($\alpha_{12} \times \alpha_{21} < 1$), and (ii) differences among species in their average performance in a given place (represented by K) must be sufficiently small so as not to overwhelm factor (i). These are essentially two ways in which species can differ from each other, and in what has been called "modern coexistence theory" (HilleRisLambers et al. 2012), these have been dubbed "niche differences" and "fitness differences," respectively (Chesson 2000b). In mathematical terms, the rate of population growth when rare, r_{rare}, is a function of these two kinds of difference as well as a scaling coefficient (s) that allows them to be expressed in units of population growth rate (MacDougall et al. 2009):

$$r_{rare} = s(\text{fitness difference} + \text{niche difference}).$$

For simplicity, I have so far described models that focus largely on community dynamics in single, closed communities, where the environment is homogenous in space and time. Many other models have been developed that relax these assumptions—for example, involving environmental heterogeneity in space or time. The consequences of different amounts of dispersal between two or more local communities have been explored in models under the umbrella of what we now call "metacommunity ecology" (Leibold et al. 2004). These types of models are treated in greater detail in Chapters 5 and 6.

Mathematical models— as well as many verbal models extending their logic to specific situations— have motivated empirical studies of various kinds (see Chaps. 8–9). Gause (1934) pioneered the use of lab microcosms containing microbes or very small bodied species (e.g., paramecia, yeast) to first estimate the parameters of a particular model and then to test its predictions in independent trials (see also Vandermeer 1969, Neill 1974). Such experiments led to Gause's "competitive exclusion principle," which essentially states that, given the inevitability of some fitness differences among species (*sensu* Chesson 2000b), coexistence of two species competing for the same resource is not possible because there is no scope for niche differentiation. Extending this principle to large numbers of species that seemingly all compete for the same few resources, Hutchinson (1961) used observations of phytoplankton in lakes to declare the "paradox of the plankton."

Many studies have aimed to characterize the differences among species (e.g., associations with different abiotic environmental conditions, or differential resource partitioning) that might allow them to coexist (Siepielski and McPeek 2010). Many other studies have searched for patterns in observational data on species distributions or community composition that are expected under strong competition—the central process of interest in the 1960s and 1970s (Diamond 1975, Weiher and Keddy 2001). One such pattern is a "checkerboard" formed by the distributions of two species, in which one or the other occurs often in any given site, but rarely the two together (Diamond 1975). Still other studies have experimentally manipulated particular factors of interest (e.g., the density or presence of other species, resource supply, dispersal) and tested whether the results reveal strong species interactions of one kind or another (e.g., competition, predation, or facilitation) or changes in community composition predicted by theoretical models (Hairston 1989). All these lines of research are alive and well in contemporary community ecology (Morin 2011, Mittelbach 2012).

3.3. LARGE-SCALE PATTERNS AND PROCESSES

More often than not, ecological patterns, and the importance of different processes in explaining them, depend on the spatial scale of observation (Levin 1992). For example, at a small spatial scale (e.g., comparing individual ponds) maximum species diversity might be found at intermediate productivity, whereas at a larger scale (e.g., comparing watersheds) species diversity might increase steadily with increasing productivity (Chase and Leibold 2002). Many definitions of an ecological community include the criterion that the species within a community interact with one another (Strong et al. 1984, Morin 2011), which consequently places an upper limit on the spatial extent of a community. Defining where to place such a limit is rather difficult, to put it mildly (see also Chapter 2), but it's fair to say that for most kinds of organisms the scale would likely be measured in square centimeters (microbes), square meters (herbaceous plants), or hectares (small mammals) rather than square kilometers. However, the core questions of community ecology— for example, why do we find different types and numbers of species in different places/times?— are literally identical to questions asked by scientists working at larger spatial scales (e.g., among biogeographic regions). Historically, such scientists might have called themselves biogeographers, whereas today they might equally call themselves macroecologists, or just ecologists. I would call them community ecologists as well.

Explanations for large-scale community patterns (e.g., comparing different continents or biomes) do involve consideration of some processes typically assumed to be of negligible importance at smaller scales. For example, the geologic and evolutionary histories of a region play major and perhaps dominant

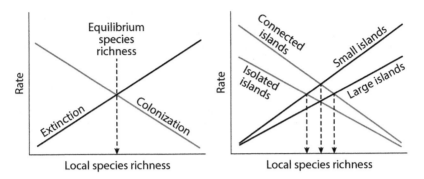

Figure 3.3. The essential features of MacArthur and Wilson's (1967) model of island biogeography, illustrating why island area and connectivity/isolation influence species richness.

roles in shaping regional biotas (Ricklefs and Schluter 1993a). However, such "regional biotas" have repeatedly come into contact and subsequently mixed, thus presenting an opportunity for "typical" community-level processes such as competition to play an important role in determining large-scale community patterns (Vermeij 2005, Tilman 2011). In addition, sharply contrasting biomes containing species with (semi)independent evolutionary histories can often occur in very close proximity (e.g., temperate forest, boreal forest, and tundra along one mountainside). Finally, the types and numbers of species contained in a regional biota or "species pool" might have an important influence on exactly how different processes are manifested as local-scale patterns (Ricklefs and Schluter 1993a), such as the relationship between species diversity and a particular environmental gradient (Taylor et al. 1990). All these observations and ideas have intellectual roots going back 100 years or more. However, their integration with small-scale studies in community ecology is comparatively more recent.

Processes thought to act at relatively large spatial scales have been represented in theoretical models in various ways. Quite in contrast with his models of locally interacting species, Robert MacArthur along with E. O. Wilson developed the "theory of island biogeography" (MacArthur and Wilson 1967), which posited that local species composition on an island was in constant flux, with species diversity determined by a balance between immigration from a continental mainland and local extinction. The resulting model predicted—and therefore helped make sense of—patterns showing reduced species diversity on smaller and more isolated islands (Fig. 3.3).

Interestingly, the key features of the island biogeography model make no important distinction between different *identities* of species (Hubbell 2001). From a pool of species on a hypothetical mainland, individuals arrive at a given rate regardless of species, and the rate of colonization (i.e., the arrival of a new species) declines with increasing local species richness because fewer and fewer of the new arrivals will represent species not already present. Larger islands can harbor larger populations, which, again regardless of species identity, have a lower chance of going locally extinct. Hubbell (2001) recognized this as one special case of a more general neutral theory, meaning a theory assuming no demographic differences among individuals of different species. He added speciation and an individual-level birth-death process to generate predictions of the shape of species-abundance distributions, species-area relationships, and the distance decay of community similarity (i.e., the decreasing similarity in the composition of communities located increasingly farther apart) at a wide range of spatial scales.

The striking match between the predictions of Hubbell's neutral theory and the empirical patterns just described caused a major controversy and a flurry of research activity in the 2000s, mostly aimed at documenting patterns not predicted by neutral theory (McGill 2003b, Dornelas et al. 2006, Rosindell et al. 2012). Many of these patterns (e.g., strong correspondence between species composition and environmental variables) were already well known. I think the longer-lasting legacy of neutral theory has been a sharp reminder that processes other than those necessarily involving species differences—specifically, drift, dispersal, and speciation—can play important roles in shaping many patterns of interest in ecological communities, regardless of whether selective processes are important in influencing some of the same patterns or even solely responsible for creating other patterns.

Speciation has long been recognized as a key factor in determining the number of species across large areas, given that it is one of only two sources of species "input" into a given area (to be discussed further in Chapter 5). In aiming to explain the latitudinal gradient in species diversity, MacArthur (1969) sketched out a model quite similar to the island biogeography model, except with a balance between immigration + speciation versus extinction, rather than only immigration versus extinction (see also Rosenzweig 1975). Indeed, it is a truism that if one area has more species than another, the balance of inputs (speciation and immigration) versus outputs (extinction) must be different.

A major push for wider recognition of the importance of regional species pools (created by speciation, immigration, and extinction) in determining the nature of local-scale community patterns came from Robert Ricklefs and colleagues in the 1980s and 90s (Ricklefs 1987, Cornell and Lawton 1992, Ricklefs and Schluter 1993a). I illustrate the basic thrust of this line of research with two examples in which predictions of hypotheses based on the dominance of local-scale species interactions contrast with predictions based on the hypothesis that properties of the regional species pool determine local patterns. First, if local species diversity is limited by competition (i.e., communities are "saturated" with species), then the number of species in small areas should not depend on the number of species in the regional pool, unless the regional

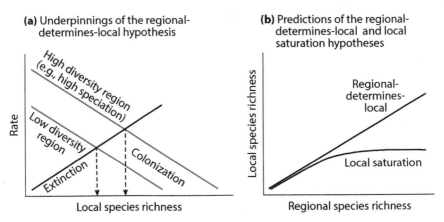

(a) Underpinnings of the regional-determines-local hypothesis

(b) Predictions of the regional-determines-local and local saturation hypotheses

Figure 3.4. Application of the island biogeography model to predicting the effect of regional diversity on local diversity (a), and a graphical depiction of competing predictions of regional versus local hypotheses. In (b) it is assumed that saturation, if present, would be manifested only above some minimal level of richness.

species pool is exceptionally depauperate. If, however, local competition is insufficiently strong to put a hard cap on the number of species, local richness should increase linearly with regional richness (Cornell and Lawton 1992). The basis of this prediction can be understood as a twist on the island biogeography model (Fox and Srivastava 2006) (Fig. 3.4a). Patterns in empirical data vary widely among systems, spanning the full range of possibilities between the two hypotheses in Figure 3.4b (see Chap. 10).

The second example pertains to explaining the shape of the relationship between species diversity and a given environmental variable (e.g., productivity). For the hump-shaped relationship often observed between species richness and productivity, a "local" hypothesis might posit that severe environmental conditions prohibit all but a few species from persisting at low productivity, severe competition reduces diversity at high productivity, and both types of species can coexist at intermediate productivity (Grime 1973). In contrast, a "regional" hypothesis might posit that competition plays no direct role but that intermediate productivity conditions have predominated over both space and time throughout the evolutionary history of the regional biota, such that more species have evolved to perform best under these conditions (Taylor et al. 1990). Thus, the effective size of the regional species pool varies among habitats with different productivities and consequently determines local diversity patterns. Testing these competing predictions with just one pattern is impossible, but if the shape of such relationships varies among regions, then according to the regional or "species-pool" hypothesis, we should be able to predict the direction of diversity-environment relationships based on knowledge of conditions that

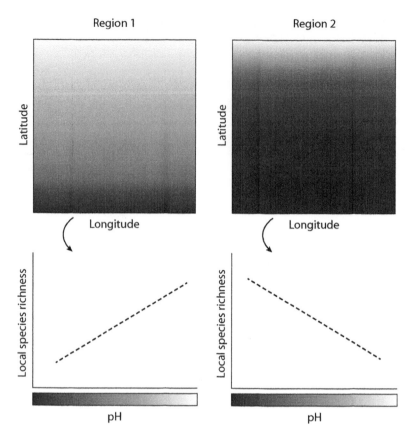

Figure 3.5. Illustration of the "species-pool hypothesis" to explain the shape of diversity-environment relationships. In region 1, high-pH conditions predominate in the region (top panel), so a positive diversity-pH relationship is found, and vice versa for region 2.

have predominated over large spatial/temporal scales (Pärtel et al. 1996, Zobel 1997, Pärtel 2002) (Fig. 3.5). Very few studies have directly tested this prediction, but they do support the species-pool hypothesis (see Chap. 10).

3.4. A SEQUENCE OF ACTIONS AND REACTIONS OVER THE LAST 50 YEARS OF COMMUNITY ECOLOGY

I think that the last 50 years of community ecology can be understood largely as a sequence of overlapping waves of enthusiasm for a particular phenomenon, process, or approach whose importance was perceived as underappreciated or understudied at a given moment in time (Fig. 3.6; see also McIntosh 1987,

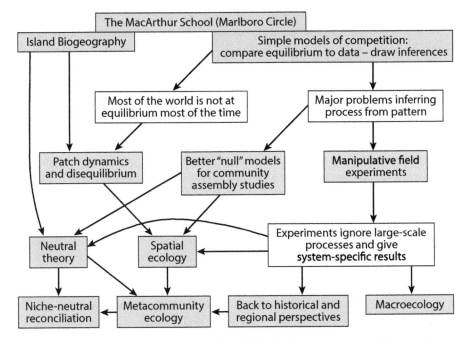

Figure 3.6. Major research programs, methods, theories, or conceptual frameworks (gray boxes) in community ecology over the last 50 years, linked by perceived weaknesses in a particular reigning paradigm (white boxes).

Kingsland 1995). Many such waves were marked by the publication of one or more books that now serve as sign posts to this history (Cody and Diamond 1975, Tilman 1982, Strong et al. 1984, Diamond and Case 1986, Ricklefs and Schluter 1993a, Hubbell 2001, Chase and Leibold 2003, Holyoak et al. 2005).

Of the three research traditions discussed in Sections 3.1–3.3, two of them gained a major thrust of momentum via the work of Robert MacArthur and colleagues in the 1960s. These colleagues included Richard Lewontin, E. O. Wilson, and Richard Levins, and collectively this group has been called the "Marlboro Circle," after Marlboro, Vermont, where they met for discussions at MacArthur's lakeside home (Odenbaugh 2013). In many ways, the other major research tradition described earlier (making sense of observational patterns) has also been largely repurposed for testing theoretical ideas that were formulated or at least clarified in this same period. The fact that dueling hypotheses (e.g., local vs. regional controls on community patterns) can both trace their origins to the same author (Fig. 3.6) has been dubbed "MacArthur's paradox" (Schoener 1983b, Loreau and Mouquet 1999). The 1960s thus serves as a good starting point for tracing the more recent origins of present-day topics of active research in community ecology.

Models based on interspecific competition as a dominant structuring force in ecological communities constituted the first wave (Cody and Diamond 1975). A great hope was that such models would provide the basis for a general and universally applicable theory of ecology (Diamond and Case 1986). This was not to be (McIntosh 1987). First, many communities are structured more strongly by predation than competition, and major criticism arose charging that any and every bit of data gathered was being interpreted as support for the competition-centric worldview without a rigorous consideration of alternative hypotheses (Strong et al. 1984). A second criticism was that the real world should not be expected to look like the equilibrium solution to a simple model, because the real world is rarely at equilibrium or simple (Pickett and White 1985, Huston 1994). These two criticisms led to the initiation or at least revival of at least three lines of research (the next waves in the sequence): (i) the use of null models to explicitly evaluate the likelihood that certain patterns might arise in the absence of competition (Gotelli and Graves 1996), (ii) a focus on perturbations from equilibrium and "patch dynamics" via disturbance (Pickett and White 1985), and (iii) the use of field experiments to test for the mechanisms underlying community patterns (Hairston 1989).

Ecologists active in the 1980s have recounted to me that it was difficult to get a paper accepted in a good journal if the study wasn't experimental. Field experiments are critical tools for testing process-based hypotheses. However, they come with severe logistic constraints in that all but a few are done at very small spatial scales (e.g., square-meter cages on a rocky shore, or plots in a grassland), and experiments are either logistically impossible or unethical in many systems (Brown 1995, Maurer 1999). Recognition of the limitations of the intense focus on processes at a local scale itself led to a new research wave, already described previously: the integration of regional processes into our understanding of communities, even at a local scale (Ricklefs 1987, Ricklefs and Schluter 1993a). One of the key processes emphasized by proponents of regional-scale phenomena is dispersal. Dispersal was already a key feature of island biogeography theory, which forces one to think explicitly about spatial scale. In the 1990s, space had been described as the "next frontier" in ecology (Kareiva 1994), and "spatial ecology" (Tilman and Kareiva 1997) was a buzz-word for a time, now manifested—at the community level—as metacommunity ecology (Leibold et al. 2004, Holyoak et al. 2005).

The theory of island biogeography (MacArthur and Wilson 1967) has had a major influence in ecology and perhaps even more so in conservation biology as a basis for predicting extinction with habitat loss and in the design of nature reserves (Losos and Ricklefs 2009). The idea that landscapes are patchy (often with island-like habitat remnants), with frequent local extinctions and colonizations, became a center piece of research under the heading "patch dynamics" (Pickett and White 1985). As described earlier, one of the more controversial theories put forward during the last 30 years in ecology—Hubbell's (2001) neutral theory—

was inspired in part by island biogeography theory. Neutral theory is considered one pillar of the metacommunity framework, and a major focus of the past 15 years or so has been an effort to reconcile the success of neutral theory in predicting some patterns in nature with the fact that one of its assumptions (demographic equivalence of individuals of different species) is clearly false (Gewin 2006, Gravel et al. 2006, Holyoak and Loreau 2006, Leibold and McPeek 2006).

3.5. PROLIFERATION AND DISTILLATION OF THEORETICAL IDEAS IN COMMUNITY ECOLOGY

With the waxing and waning of various models, conceptual ideas, buzzwords, methods, and philosophies in community ecology over the past century, a student could be forgiven for finding difficulty in seeing any kind of overarching structure into which everything fits. Each new perspective or theory has typically emphasized one or a few processes—not necessarily to the exclusion of others, but at least with a focus on a particular subset: neutral theory emphasizes everything except selection, niche theory focuses on selection, metacommunity theory emphasizes dispersal, and so on. The various waves of interest in different topics during the time that I have been a student of community ecology ultimately laid bare (to me) the fact that all the processes of interest can be reduced to four analogues of the processes in population genetics, which students have no trouble easily seeing as the overarching conceptual structure into which everything in that discipline can fit.

In communities, the term *selection* has been used only sporadically to describe a process acting among individuals of different species (Loreau and Hector 2001, Norberg et al. 2001, Fox et al. 2010, Shipley 2010), but all the deterministic outcomes of ecological models involving differences between species, from Lotka-Volterra to the present, are essentially models of selection in communities (Vellend 2010). So, selection has always been a conceptual focus of community ecology. The potential influence of community drift, via demographic stochasticity, has been recognized for a long time, but it took root in the field as a whole only after Hubbell (2001) stirred the pot with his neutral theory. Likewise, dispersal has featured in prominent ecological models for many decades, but development of the metacommunity concept (Leibold et al. 2004) has served as a reminder of its central place as a distinct process influencing communities. Finally, the importance of considering the formation of regional species pools when studying communities at any scale (Ricklefs and Schluter 1993a), as well as the emergency of macroecology (Brown 1995), added speciation to the mix of distinct processes that can influence ecological communities. With these four processes in hand, the smorgasbord of theory in community ecology can be reined in and understood as many combinations of a few key ingredients.

PART II
THE THEORY OF ECOLOGICAL COMMUNITIES

CHAPTER 4

The Pursuit of Generality in Ecology and Evolutionary Biology

The theory of ecological communities is essentially my attempt at identifying and articulating the most general theoretical principles that cut across the plethora of ideas, concepts, hypotheses, and models in community ecology. Before I describe the theory in detail (Chap. 5), it is thus useful to consider other ways in which community ecologists have sought generality, why the results have been less than perfectly satisfactory, and why the body of theory from which I have taken inspiration—population genetics—is so widely accepted as providing a general and robust theoretical foundation for evolutionary biology. These are the goals I pursue in the present chapter, in addition to providing a brief summary of the theory of ecological communities.

4.1. GENERAL (AND NOT-SO GENERAL) PATTERNS IN ECOLOGICAL COMMUNITIES

A dominant theme in ecology for many decades has been the search for patterns in nature that are "general," meaning those that appear in a similar form under many different circumstances (e.g., not just in coral reefs, in arctic tundra, or in tropical forests, but in all three). MacArthur (1972) famously said that "to do science is to search for repeated patterns, not simply to accumulate facts." We can think of this as a "pattern-first" approach to seeking generality in the science of ecology (Cooper 2003, Roughgarden 2009, Vellend 2010), and indeed, many repeated patterns can be found in nature. For example, the number of species almost always increases as a function of area, and such species-area relationships take a restricted range of forms (Rosenzweig 1995); communities typically contain a few abundant species and many rare ones (McGill et al. 2007); and for many different groups of organisms, species diversity is greatest in the tropics and declines steadily toward the poles (Rosenzweig 1995). However, as far as I know, literally none of the "general" patterns in community ecology are universal (e.g., see Wardle et al. (1997) for a negative species-area relationship). Some relationships initially considered to be highly general have subsequently been found to vary greatly among systems (e.g., the diversity-productivity relationship; Waide et al. 1999). And most important, while one motivation for the pattern-first approach is the idea that a general pattern suggests a generally applicable cause or process (Brown 1995), many patterns can, in fact, be generated by several different causal pathways, often referred to as the "many-to-one" problem (Levins and Lewontin 1980). Consequently, the identification of frequently observed patterns in nature has not resulted in a general theory of ecological communities.

4.2. THE PROCESSES THAT UNDERLIE PATTERNS IN ECOLOGICAL COMMUNITIES

Another way to seek generality in community ecology is to lay out a set of processes thought to underlie whatever patterns we might observe in nature (Shrader-Frechette and McCoy 1993). This is a "process-first" approach, and the question is, what processes or mechanisms can cause community properties to change over space and time? The answer can be sought in various ways.

The typical set of factors or processes used to explain community patterns includes dispersal, which determines the set of species that has access to a site; abiotic factors such as climate or disturbance; and biotic factors such as competition and predation. These factors and processes are often conceived of as a series of filters determining which species from a regional pool of candidates is observed in any given site (Keddy 2001, Morin 2011, Fig. 4.1). One can think of a very long list of specific factors that might play a role in this filtering process, and an even longer list of ways such factors might interact, thus providing the intellectual space for innumerable theories, models, or conceptual frameworks emphasizing particular factors of interest.

An important empirical lesson derived from thousands of case studies is that the influence of any one factor on community composition is highly contingent on system-specific details (Lawton 1999). For example, sometimes, grazing or climate warming will increase plant diversity, and sometimes it will do the opposite (Vellend et al. 2013). Sometimes, removing a predator will cause massive changes in the rest of the community, and sometimes it won't (Shurin et al. 2002). Progress in understanding can be advanced, then, by determining the conditions under which a given factor will have this or that effect, by ex-

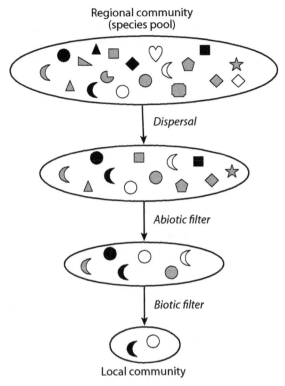

Regional community
(species pool)

Dispersal

Abiotic filter

Biotic filter

Local community

Figure 4.1. The filter model of community assembly. In this hypothetical example, a random subset of species in the regional pool has access to a local site; only species with rounded edges can tolerate the abiotic conditions; and competition leads to the elimination of all but one species of each functional type (shape). The two remaining species can coexist stably given contrasting resource requirements.

amining how different driving factors interact. However, the extreme system specificity in results sought at this level makes the pursuit of generally applicable theory or the identification of ecological laws a distant dream at best and a certain dead end at worst (Beatty 1995).

The debate over whether the pursuit of ecological laws should be a central goal of our science, or whether it should be abandoned, has persisted for a long time (MacArthur 1972, McIntosh 1987, Shrader-Frechette and McCoy 1993, Lawton 1999, Cooper 2003, Simberloff 2004, Scheiner and Willig 2011). Importantly, the terms of the debate typically involve an implicit assumption that processes must be considered in the way I have just described—that is, a general theory of ecological communities must involve factors like climate, disturbance, dispersal, predation, competition, and mutualistic interactions. However, one can think about ecological processes at different levels. For example, the processes that determine changes over space and time in allele frequencies in populations are largely ecological in nature (mutation is the only exception), and as described in the next section, population geneticists have approached things quite differently than ecologists.

4.3. THE THEORY OF POPULATION GENETICS: HIGH-LEVEL PROCESSES

The central question that ultimately led to the conceptual framework proposed in this book is as follows: why have evolutionary biologists achieved near unanimity that their discipline is underlain by a rock-solid and highly general theoretical foundation (population genetics), while ecology is widely perceived as being a theoretical and conceptual basket case? Van Valen and Pitelka (1974) put it succinctly: "Unlike population genetics, ecology has no known underlying regularities in its basic processes."

Population genetics and community ecology are both concerned with understanding how the composition and diversity of biological variants vary across space and time. From a theoretical standpoint, whether these variants are alleles or genotypes (population genetics) or species (community ecology) is often irrelevant. In evolutionary biology, the reconciliation of Darwin's theory of evolution by natural selection and Mendel's demonstration of particulate inheritance occurred under the banner of the *modern evolutionary synthesis* (Mayr 1982, Kutschera and Niklas 2004). One important outcome of this synthesis was a coherent set of population-genetic models describing evolutionary change as the outcome of four key processes: selection, drift, migration (gene flow), and mutation (Hartl and Clark 1997). This set of models provides the core foundation from which the law-like status of evolutionary theory arose (notwithstanding contemporary debate as to the sufficiency of this theory; Laland et al. 2014).

So, why doesn't community ecology have a coherent set of interconnected models united under a conceptual framework akin to the modern evolutionary synthesis? In some ways, I think it already does, but it simply hasn't been recognized as such given a failure to consistently distinguish between what we can think of as "high-level" processes (defined in the next section) from "low-level" processes. In evolutionary theory, natural selection is one of four high-level processes (along with drift, gene flow, and mutation). If selection is of constant magnitude, one allele will be favored over the others, and it will increase in frequency to the eventual point of fixation. If the fitness of a given allele depends on its frequency, selection can favor the stable maintenance of multiple alleles (negative frequency dependence), or it can make the "winning" allele dependent on initial conditions (positive frequency dependence). Note that these statements say nothing about *why* one allele has higher fitness than the others. The reasons might involve resource competition, tolerance of environmental stressors, avoidance of predators, or any number of other factors that confer success under particular conditions. I refer to all these reasons as "low-level processes." Darwin (1859) lumped them under instances of the "struggle for existence," and Ernst Haeckel (translated in Stauffer 1957) simply called them "ecological." In essence, I view a horizontal ecological community as a "population" in which the organisms belong to multiple species instead of a single species. Selection among individuals of different species in a community is then conceptually identical with selection among individuals of different asexual genotypes in a single-species population. (I elaborate on this idea in Chap. 5.)

As argued by Elliott Sober (Sober 1991, 2000), the models of natural selection that have universal generality are those that address the *consequences* of fitness differences among individuals with different genotypes, rather than the *causes* of those fitness differences. The causes of fitness differences are innumerable: a genotype might have relatively high fitness because it is a dark-colored moth invisible to predators on trees that have recently been blackened by pollution (Kettlewell 1961), because it is a bird with a large beak in a year when most available seeds to eat are large (Grant and Grant 2002), or for any number of other reasons. However, the expected consequences of these fitness differences for evolutionary dynamics are the same in each case. The mean trait value should shift toward that favored by selection. As these examples illustrate, theories or models that address the causes of fitness differences are potentially infinite in number, with the applicability of each one highly contingent on any number of factors (Beatty 1995, Sober 2000). Thus, the strength of evolutionary theory lies in its focus on high-level processes: selection, drift, migration, and mutation. In other words, "Evolutionary theory achieves its greatest generality when it ignores sources and focuses on consequences" (Sober 1991). Theory in community ecology has focused on both high-level and low-level processes, but this distinction has not been emphasized, and collectively the number of competing models one might build with low-level processes is overwhelmingly large.

4.4. HIGH-LEVEL AND LOW-LEVEL PROCESSES IN COMMUNITY ECOLOGY

As with (micro)evolutionary change, community dynamics is underlain by just four high-level processes: selection, drift, dispersal, and speciation (Vellend 2010). The framework resulting from this perspective will be developed further in Chapter 5. As an initial brief illustration, consider one of the simplest questions one can ask about an ecological community: what can cause a change in the number of species (S) present in a particular place over time? Figure 4.2 shows one community, initially with nine individuals divided among three species. There are only four fundamental ways to change the number of species, which we can imagine in a hypothetical sequence of steps:

1. *Speciation*: One subpopulation of the white species diverges from the other, forming a new species; the community now has four species instead of three.

2. *Dispersal* (immigration): An individual of a species not already present locally arrives from somewhere else and joins the community. A species has been added to the community.

3. *Drift*: Even if all individual organisms have the same expected rates of survival and reproduction, there is a nonzero probability that, for example, the solid gray individual will die before reproducing. If this happens, the number of species will decrease.

4. *Selection*: The white species has a fitness advantage over the black and dotted species (i.e., it deterministically makes a greater per capita contribution to the community in the next time step), thereby excluding them from the community. The number of species has decreased again.

To summarize in mathematical terms, $S_{t+1} = S_t$ + speciation + immigration – extinction (see also Ricklefs and Schluter 1993a), with two different pathways to local extinction: drift and selection. (Note that while speciation, immigration, and extinction might appear to represent a smaller and more fundamental set of processes than the four I propose, they help directly explain only variation in species numbers, not composition or abundance, and thus do not encompass a comprehensive set needed for ecological communities.) If one adds different forms of selection to these scenarios—slowing, preventing, or accelerating extinction—the result is a logically complete description of high-level processes that can influence the dynamics not only of the number of species but of any community property one might wish to examine. The only exception would be a community property that incorporates variation within species (e.g., for traits), which I will largely ignore to keep things as simple as possible.

Ecologists, myself included, love to emphasize the complexity of their subject matter. So many things are going on simultaneously in an ecological community that the number of ways one can carve out a research niche is limited only by the imagination of the researcher. Hot topics over the years have included the importance of competition versus predation, equilibrium versus nonequilibrium species coexistence, omnivory, nonconsumptive effects of predators, plant-soil feedbacks, facilitation (or positive interactions in general), the stoichiometry of organisms and their environments, and the interactive effects of

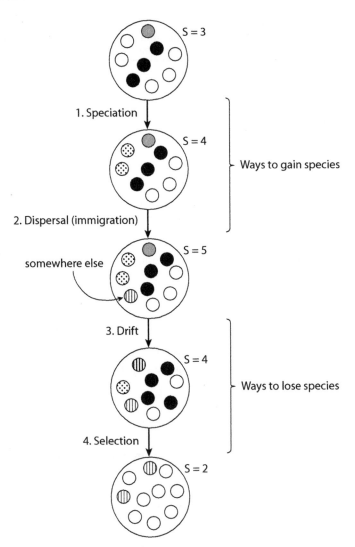

Figure 4.2. The four high-level processes that can cause the number of species (S) in a community to change. Each small circle is an individual organism, with the fill type indicating different species.

climate and species interactions. These topics are distinguished from one another largely by the different *causes* of fitness differences among species on which they focus. In other words, they address low-level processes that can lead to a high-level agent of change: selection. Viewed in this light, Sober's (1991, 2000) argument about where the generality lies in the theory of evolu-

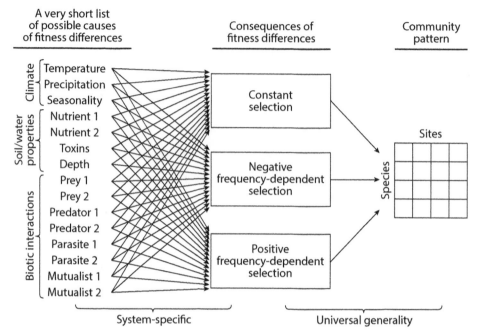

Figure 4.3. The causes of fitness differences between species are innumerable and system specific, whereas the possible consequences of fitness differences are relatively small in number and universally applicable.

tion by natural selection seems especially cogent in the context of community ecology. Focusing only on selection for the moment, we know that the importance of different causes of selection varies tremendously from case to case and is contingent on an overwhelmingly long list of potential factors (Lawton 1999). This makes the hope for general theory in community ecology starting from low-level processes quite faint. The *consequences* of different forms of selection, however, are universally applicable and therefore a promising target for the development of general theory (Fig. 4.3). An illustration of the power of viewing things this way is Chesson's (2000b) distillation of the many models of stable species coexistence, which may involve any number of low-level processes, to two key ingredients: negative frequency-dependent selection and constant selection. Chesson refers to these as "stabilizing mechanisms" via "niche differences" and "equalizing mechanisms" via small "fitness differences," respectively. I elaborate on this and other perspectives in the next chapter.

Returning to the question of complexity in ecological communities, Allen and Hoekstra (1992) argue compellingly that "complexity in ecology is not so much a matter of what occurs in nature as it is a consequence of how we choose to describe ecological situations. . . . it is decisions of the observer that make a system complex." So, one can choose to view ecological communities as the product of a highly complex set of low-level processes or as the product of a fairly simple set of high-level processes. I think that a focus on low-level processes is entirely appropriate and indeed critically important when aiming to understand how particular systems work, and in applied settings when particular drivers are of interest (Shrader-Frechette and McCoy 1993, Simberloff 2004). However, if the aim is conceptual synthesis or the development of general theory in community ecology, I think that high-level processes may be the only place that either is possible.

4.5. A PATHWAY TO GENERAL THEORY IN COMMUNITY ECOLOGY

Textbook treatments of a given topic reflect the way that practitioners and students of a discipline view the conceptual organization of its contents. Writers decide how to conceptually synthesize the material, and students learn the material based on the resulting organizational structure. In my experience (and in the experience of many colleagues), students find the conceptual organization of community ecology confusing. The same is true of ecology more broadly (D'Avanzo 2008, Knapp and D'Avanzo 2010). I think one of the reasons is that the overarching themes are a grab-bag of unlike entities.

The left side of Figure 4.4 shows a simplified list of the major organizing themes in a typical treatment of community ecology (Putnam 1993, Morin 2011, Mittelbach 2012). These represent the *patterns* of interest, selected low-level *processes* (competition, predation), some *abstract concepts* (niches, food webs), and an *important thing* always to keep in mind (scale). These themes, along with the many subthemes in such treatments, reflect quite well the way that researchers have carved up the research landscape. Along the way students will learn of many repeated patterns (e.g., species-area relationships, relative

Community Ecology Textbook		**Population Genetics Textbook**	
Topic	Type of entity	Topic	Type of entity
Ecological patterns	Pattern	Genetic patterns	Pattern
Competition	Low-level process	Selection	High-level process
Predation	Low-level process	Drift	High-level process
Food webs	Abstract concept	Migration	High-level process
Niches	Abstract concept	Mutation	High-level process
Scale	Thing to think about		

Figure 4.4. Organizing themes in textbook treatments of community ecology and population genetics. Modified from Vellend and Orrock (2009).

abundance distributions) and of a great many low-level processes at work in ecological communities that might produce such patterns. This is why, as mentioned earlier, if each student in an undergraduate or graduate class is asked to write down a list of processes that can influence community composition and diversity, the result will be a long list from each student, with substantial differences among lists. This is the cost of organizing the material of community ecology based on low-level processes, observational patterns, and an assortment of cross-cutting abstract concepts. It may also present a constraint on research progress in the field to the extent that it prevents a common view of what exactly the discipline is all about (Shrader-Frechette and McCoy 1993, Knapp and D'Avanzo 2010, van der Valk 2011).

An alternative approach is to start, as previously, by describing the patterns of interest but to subsequently focus on high-level processes as the organizational theme (the right side of Fig. 4.4). I hope to demonstrate in subsequent chapters that almost all theoretical models and ideas in community ecology can be understood in relation to one another based on their emphasis on four high-level processes—selection, drift, dispersal, and speciation—and a manageable set of ways these processes are manifested (e.g., forms of selection) and interact. In essence, the high-level processes provide a key missing link that can help simplify our understanding of how low-level processes result in the dynamics and patterns we observe in ecological communities.

The generalized set of models and ideas that describe community dynamics and structure based on the action of high-level processes can be thought of collectively as the theory of ecological communities. I am not claiming to have devised a new general theory from scratch, or even to have developed any of its component models. Rather, I have imported the core ideas of an existing theory (population genetics) and conceptually organized the existing material in community ecology in a way that allows one to filter the complex details through a (relatively) simple set of high-level and universally applicable processes that determine the dynamics and structure of ecological communities.

CHAPTER 5

High-Level Processes in
Ecological Communities

Over a given interval of time, an individual organism can grow, reproduce, move somewhere else, or die. The probabilities of these outcomes might vary among individuals of different species in a community, or they might not, and from time to time a new species might arise via speciation and join the community. The sum total of these events creates the patterns we observe over time and space in ecological communities. As described in Chapter 4, only four high-level processes can underlie community dynamics: drift, selection, dispersal, and speciation. From the point of view of one community considered at any scale of space or time, species can be added via speciation or dispersal (immigration). The actions of selection, drift, and ongoing dispersal subsequently shape the abundances of these species, possibly pushing some to local extinction. That, in a nutshell, is the theory of ecological communities.

   In this chapter I first elaborate on the nature of the four high-level processes in the general theory. I have attempted to minimize repetition of material from the previous chapters and with an earlier paper outlining the theory (Vellend 2010), but some rehashing is necessary to keep things coherent. I pay extra attention to trait-based selection and to speciation, as these were least developed in Vellend (2010). At the end of the chapter (Sec. 5.7) I demonstrate how each theory in a fairly long list in community ecology can be understood with reference to a particular combination of high-level processes.

5.1. THE GENERAL THEORY

"Changing the domain of a model can be a fruitful avenue for juxtaposing phenomena or processes that had been considered in isolation" (Scheiner and Willig 2011, 5). The theory presented here juxtaposes theories of population genetics, parts of quantitative genetics, and of community ecology. However, the core ideas of these theories are actually applicable more broadly, to "all self-replicating things" (Bell 2008, 16; see also Lewontin 1970, Nowak 2006). These "things" might be individual organisms of one species, individuals of different species, strings of computer code ("digital organisms"), or a particular kind of business practice in a human society (Mesoudi 2011). Darwin's own definition of natural selection could just as easily apply to multispecies communities as to single-species populations: "This preservation of favourable variations and the rejection of injurious variations, I call Natural Selection" (Darwin 1859).

The aggregate properties of a collection of self-replicating things (e.g., a mean trait value in population, the number of species in a community, the dominance of a particular business practice) arise via four fundamental processes: origination of new types, movement from place to place, stochastic sampling from one time point to the next, and selection. In population genetics, these properties are represented by mutation, migration (gene flow), genetic drift, and natural selection, respectively (Hartl and Clark 1997). In community ecology, they are represented by speciation, dispersal, ecological drift, and selection (Vellend 2010). In both cases, we assume that individuals are engaged in a "struggle for existence" (Darwin 1859, Gause 1934) of some kind. Rather than meaning competition in the narrow sense of a direct negative effect of one type on another, the idea of a "struggle for existence" simply recognizes that there are limits to the growth in numbers of any one type.

The influence of each of the four processes, and interactions among them, can be understood with reference to how they influence patterns of interest in ecological communities. The following sections describe the basic principles, with the hope they are intuitive. Chapter 6 illustrates these principles using quantitative simulation models and provides readers the tools to explore such simulations on their own.

5.2. ECOLOGICAL DRIFT

Life, death, reproduction, and dispersal all involve probabilities (McShea and Brandon 2010). Individual fitness is a product of these probabilities and can be conceived of as follows: "Trait X is fitter than trait Y if and only if X has a higher probability of survival and/or a greater expectation of reproductive success than Y" (Sober 2000). Thus, knowing that an organism possesses trait

X—or in the case of community ecology, knowing an individual's species identity—might allow us to predict the probabilities of different demographic outcomes, but this knowledge does not allow us to predict the fate of any particular individual with certainty (Nowak 2006).

If, all else being equal, individuals of species 1 and species 2 have survival probabilities of 0.5 and 0.4, respectively, then we *expect* species 1 to rise to dominance. That is, species 1 has greater fitness. However, imagine that there are just two individuals of each species and that survival is determined prior to reproduction (we will assume that the latter is asexual). The probability that both individuals of species 1 die before reproducing is not so small: $(1 - 0.5)$ $\times (1 - 0.5) = 0.25$. At the same time, there is a good chance that at least one individual of species 2 will go on to reproduce, which can be calculated as one minus the probability that both individuals of species 2 die: $1 - (0.6 \times 0.6) = 0.64$. So, despite the higher fitness of species 1, species 2 can "win" because there is a random component to community dynamics (i.e., the probabilities). This random component is called *ecological drift*.

If a community contains many individuals—let's say 1000 each of two species —then after an episode of mortality, the number of remaining individuals of each species will almost certainly be very close to $0.5 \times 1000 = 500$ for species 1, and $0.4 \times 1000 = 400$ for species 2. For species 1, it's equivalent to flipping a coin 1000 times: for any individual flip, all we can do is guess at the outcome, but since each flip is independent, we can be quite certain that after 1000 flips the heads and tails will roughly balance out. The odds of either <400 or >600 heads is vanishingly small (Fig. 5.1b). However, if we flip the coin only 10 times (i.e., if the population size is only 10), we can say that five heads and five tails is the most likely outcome, but there is a decent chance it will be some other outcome (Fig. 5.1a).

In the hypothetical example just described, one of the two species had a fitness advantage, and so selection was part of the scenario. I started with this scenario deliberately to illustrate that ecological drift does not require an assumption of neutrality—that is, an assumption that individuals of different species are precisely equivalent demographically. However, the influence of drift on community dynamics is greatest and easiest to think through in the purely neutral scenario, simulations of which will be explored in Chapter 6. Briefly, assuming some upper limit on total community size, drift in a local community causes species relative abundances to fluctuate at random, irrespective of their identities or traits. Eventually, one species will drift to complete dominance ("fixation" in population genetics), with all other species drifting to extinction. Species initially at low frequency are those most likely to drift to extinction, because, for example, random fluctuations are more likely to reach zero (a stable end point) if initial frequency = 0.05 than if initial frequency = 0.95. In mathematical terms, the probability of a species reaching complete dominance (the only other stable end point) is equal to its initial frequency (Hubbell 2001).

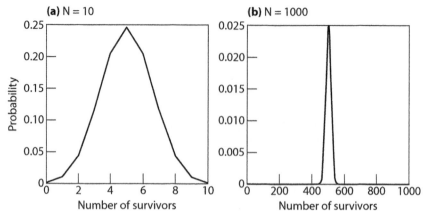

Figure 5.1. The crux of ecological drift. With a probability of survival of 0.5, the number of actual survivors can be quite variable if the initial population is small (a), but much more predictable if the population is large (b).

Maintaining diversity in a purely neutral community thus depends on a source of input of new species, which can happen via immigration or speciation. In multiple habitat patches that are not linked by dispersal, species abundances drift randomly and independently within each local community, thus causing their species composition to diverge (i.e., beta diversity increases).

Before moving on from the topic of drift, it is worthwhile clarifying what exactly is meant by "random," or equivalently, "stochastic." For centuries, scientists and philosophers of science have grappled with the question of whether anything in the universe truly happens at random (Gigerenzer et al. 1989). One point of view is that nothing is genuinely random, such that our view of demography (or anything else) as probabilistic is simply a reflection of our inability to predict (i.e., a lack of knowledge) rather than a reflection of anything being truly unpredictable (Clark 2009, 2012). However, within a defined domain of interest, it is theoretically possible for some events (e.g., birth or death) to occur at random *with respect to* the identities of each individual object (e.g., the species identity of an organism) (Vellend et al. 2014). This "with respect to" conception of stochasticity was articulated by McShea and Brandon (2010), and it offers a pathway to operational definitions of "true" stochasticity in particular settings. In population genetics, drift at a particular locus occurs when demographic events occur independently of the allele(s) an individual carries at that locus (Hartl and Clark 1997). This is not controversial. That each demographic event has a nonrandom cause (e.g., death by disease) is irrelevant; what's important is that many such events occur randomly *with respect to* allelic states of a particular locus. Likewise, we can say that "true" ecological drift occurs in a community when demographic events occur at random *with respect to* species identities (Vellend et al. 2014). This concept should be no more controversial than the theoretical idea of genetic drift.

5.3. SELECTION

Selection in ecological communities is at the same time the most familiar of processes and also the one whose name causes the most confusion. Many biologists have become so accustomed to thinking about selection only as a process underlying genetic evolution within species that we have lost sight of the fact that it is a process with much more general application, and has been considered so right from its inception (Darwin 1859, Lewontin 1970, Levin 1998, Loreau and Hector 2001, Norberg et al. 2001, Fox et al. 2010, Mesoudi 2011). Bell (2008) has described selection as the key process underlying one of only two systems of knowledge needed to understand the natural world. Physical laws underlie the "first science," which includes physics, chemistry, and physiology, among others. The "second science," which includes evolutionary biology and community ecology, as well as economics, involves "the operation of selection on variable populations, and . . . cannot be understood in terms of nor reduced to the principles of physics and chemistry" (Bell 2008).

In an ecological community, selection results from deterministic fitness differences between individuals of different species (Vellend 2010). But what is fitness? In short, fitness can be defined in almost exactly the same way as in evolutionary models of asexual, single-species populations, just by substituting the word *species* for *genotype*. Absolute individual fitness is the expected quantity of offspring produced by an organism per unit time, including survival of the organism itself. To predict community dynamics, we need to first calculate the absolute fitness averaged across individuals within each species in the community. The *relative* fitness of each species is the absolute fitness standardized in some way across species, for example, by dividing by the community-level average or by the absolute fitness of the fittest species (Orr 2009). Selection is expected to result in a change in community composition (i.e., the relative abundances of species) to the extent that species vary in their average relative fitness.

Before proceeding to different forms of selection, it is worth noting three integral issues in the previous paragraph. First, both the first and last sentences include the word "expected." Why is that? As shown in the previous section on ecological drift, the vagaries of stochastic birth and death mean that even two genetically identical individuals in the same environment might not leave the same number of offspring. To repeat an important quote: "Trait X is fitter than trait Y if and only if X has a higher probability of survival and/or a greater expectation of reproductive success than Y" (Sober 2000). If neither of our two hypothetical individuals has a higher *probability* of survival or a greater *expectation* of reproductive success, there is no fitness difference, even if one individual, by chance, leaves more offspring than the other. My definition of selection thus refers to fitness differences as "deterministic." Philosophers have tied themselves in knots over this issue (Beatty 1984, Sober 2000, McShea and Brandon 2010), although for the purposes of this book we need not pursue it further.

The second issue concerns the expression "quantity of offspring" instead of "number of offspring," the latter being the typical phrasing in evolutionary biology (Orr 2009). This choice is driven by the fact that species "abundances" are very frequently quantified not by counting individuals but by estimating factors like biomass, biovolume, or percent ground cover. This is because for some types of organisms, plants especially, delineating individuals can be rather arbitrary. For example, is a grass "individual" a single stem, a cluster of stems, a group of stem clusters physically connected by rhizomes, or all the stems in a genetic clone even if physical connections have been broken? There is no one good answer to this question. Measures such as biomass also allow accounting for differences in body size among species (e.g., one moose can count for more units of "abundance" than one rabbit). In sum, fitness in ecological communities need not be expressed in terms of numbers of offspring, reflecting both biological reality and a need to make the concept empirically operational.

The final issue worth noting is the absence of the concept of heritability from my discussion of selection and community dynamics. In essence, I assume perfect heritability in the sense that the offspring of a moose will always be another moose, and the offspring of a rabbit will always be another rabbit. Notwithstanding hybridization between species and hotly debated species concepts (Coyne and Orr 2004), essentially all of community ecology implicitly assumes the same thing when binning individuals into discrete taxa (most often species) and then assessing changes in abundances within those bins across space and time.

5.3.1. Forms of Selection

Selection appears in several important forms, depending on whether its direction and strength vary according to the current state of the community itself, or over space and time (Nowak 2006). These scenarios can be visualized by first imagining a two-species community in which we are interested in relative rather than absolute abundances. Since the relative abundance (i.e., frequency) of species 2 is just one minus the frequency of species 1, the state of the community is fully specified by calculating the frequency of species 1 (the x-axis in Fig. 5.2). With only two species, at a given moment in time, selection can favor only one or the other, such that the fitness difference between the two species (the y-axis in Fig. 5.2) alone determines the expected change in frequency.

Throughout the remainder of the next two chapters, the discussion focuses on species frequencies (i.e., relative abundances) rather than absolute abundances or densities, despite the fact that most theoretical models in ecology focus on the latter, while population genetic models focus on the former (Lewontin 2004). In a community of constant size, the two are equivalent. While the mechanistic details connecting low-level processes to the outcome of species interactions may often involve absolute densities (Chase and Leibold 2003), assuming a constant total abundance across species can be a reasonable approximation (Ernest et al. 2008). In addition, community-level patterns are almost all quantified based on relative abundances, and the latter also permit illustration of the essential features of the link between high-level processes and community patterns. For example, frequency dependence rather than density dependence is the key to understanding species coexistence (Adler et al. 2007, Levine et al. 2008).

The simplest kind of selection is constant with respect to frequencies of species, favoring one over the other, independent of the state of the community itself (Fig. 5.2a). This situation might occur, for example, if extrinsic environmental conditions (e.g., climate) cause greater fitness of one species relative to the other. The expected end result is dominance by one species and thus a reduction in local diversity.

If the strength or direction of selection depends on the state of the community itself, there is potential for negative or positive feedbacks to arise. If, for example, individual fitness in each of two species is most strongly limited by a different resource (e.g., nitrogen or phosphorus), then when a given species is relatively rare, its limiting resource should be abundant, and so its population growth rate should be relatively high (Tilman 1982). This situation represents negative frequency-dependent selection, and if it is of sufficient strength, both species should tend to increase in abundance when rare and decrease when abundant, thus creating a stable equilibrium at which both species are present in the community—that is, species coexistence (Fig. 5.2b). Negative frequency-dependent selection that is strong relative to constant selection should help maintain relatively high levels of species diversity (Chesson 2000b).

Positive frequency dependence means that each species has a fitness advantage when common, thus making the outcome dependent on initial conditions. This situation might occur if, for example, a given plant species modifies the soil environment in a way that benefits conspecifics (e.g., promoting mutualistic fungi) and that inhibits heterospecifics (e.g., by altering soil pH) (Bever et al. 1997). In such scenarios, one species or the other is expected to dominate, with the winner determined by which species starts out at a greater abundance relative to the unstable equilibrium point at which the two species have equal fitness (Fig. 5.2c). Such "priority effects" are one simple case of multiple stable states existing for a given set of environmental conditions, which often arise via positive feedbacks of some sort (Scheffer 2009). Positive frequency-dependent

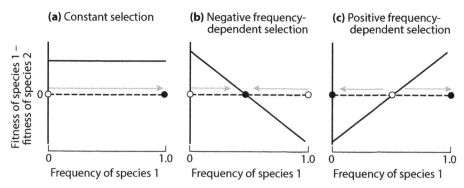

Figure 5.2. Three fundamental forms of local selection among individuals of different species. Solid lines show relationships between fitness and frequency, while the dashed horizontal line in each panel indicates zero on the y-axis. Gray arrows indicate the direction of expected community change, open circles indicate unstable equilibrium points, and filled circles indicate stable equilibrium points. The other two important forms of selection—spatially variable and temporally variable—occur when the strength and nature of selection varies among local communities or across time. These scenarios are modified from Nowak (2006) and Vellend (2010).

selection should reduce local diversity given that one species is expected to rise to dominance in each site. It should also increase beta diversity (different species will dominate in each site) and obscure composition-environment relationships (the identity of the dominant depends on initial frequencies rather than the environment).

Finally, the strength and direction of selection might vary over space (i.e., among local communities) or across time. For example, if climatic conditions vary over space or time, species 1 might be favored in some years and places (Fig. 5.2a), and species 2 in other years or places (imagine Fig. 5.2a with the solid line below zero on the y-axis). Depending on the details of selection, community size, and dispersal, spatiotemporally variable selection can, in some cases, promote coexistence and diversity (Chesson 2000b). Because spatially variable selection systematically favors different species in different places according to local environmental conditions, it should generally create and maintain strong composition-environment relationships, the result of which is relatively high beta diversity. These predictions are explored further in Chapter 6.

5.3.2. Explicitly Trait-Based Selection

The theoretical forms of selection discussed thus far in the context of ecological communities are identical to those we might expect to occur among asexual genotypes in a single-species population (Nowak 2006), as captured

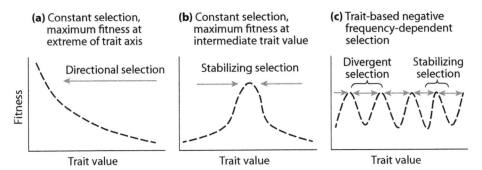

Figure 5.3. Consequences of different forms of selection for community trait composition. Gray arrows point from trait values with low fitness to trait values with high fitness. In (a) and (b), the fitness functions might be fixed based on local environmental conditions; in (c), the wavy pattern emerges as a consequence of trait-based negative frequency-dependent selection, but the positions of the peaks depend on initial conditions in a model (Scheffer and van Nes 2006). Note that these are stylized rather than quantitative versions of model predictions.

in population-genetic models (Hartl and Clark 1997). The influence of selection on community patterns has thus been discussed with reference to either first-order community properties (i.e., those not involving traits or environmental variables) or to second-order properties involving environmental variables (see Chap. 2). With the explicit incorporation of traits into the quantification of community-level patterns, the process-pattern relationship becomes analogous to that found in models of quantitative genetics, in which the focus is on changing trait distributions (Falconer and Mackay 1996).

In community ecology, the basic logic of trait-based analyses involves the following two scenarios. First, high fitness might be permitted by only a narrow range of trait values, either because such trait values confer tolerance of stressful environmental conditions or because they allow species to outcompete others with different trait values (Weiher and Keddy 1995, Mayfield and Levine 2010). With respect to community trait composition, what emerges from such constant selection is akin to either directional selection (if the favored trait value is at one extreme of the community-level trait distribution) or stabilizing selection (if the favored trait value is intermediate) (Fig. 5.3a, b). In either case, we expect a reduced range or variance of species mean trait values in the community (Weiher and Keddy 1995).

The second scenario involves negative frequency-dependent selection, with a twist. We first assume that the fitness of a given individual declines not only with increasing frequency of conspecifics (basic negative frequency dependence) but also with increasing abundance of any individuals that share similar trait values, regardless of species (trait-based negative frequency dependence).

Essentially, we assume that species with more similar trait values (e.g., beak size, rooting depth) compete more strongly. What emerges here is analogous to a combination of diversifying selection and stabilizing selection, depending on which range of trait space is considered (Fig. 5.3c). The predicted pattern is increased variance and even-spacing of trait values in the community (Weiher and Keddy 1995).

5.4. DISPERSAL

Dispersal is the movement of organisms from one place of residence to another, as distinct from seasonal animal migration, which involves systematic back-and-forth or repeated cyclical movement among the same set of places. Dispersal is one of only two ways to add new species to a community, and thus greater incoming dispersal (i.e., immigration) should lead to greater local diversity (MacArthur and Wilson 1967). Dispersal also means that local communities no longer undergo dynamics independently, so dispersal should increase the similarity in species composition across space, thereby reducing beta diversity (Wright 1940), assuming species can establish in places they arrive. Dispersal also involves emigration and therefore loss of individuals (or propagules) from a local community, although the consequences of this aspect of dispersal are not as well studied.

Following immigration of a new species to a local community, subsequent effects on species composition and diversity may occur if the new species alters the selection regime or the population sizes of resident species. These are not primary effects of dispersal per se but secondary effects via a change in community composition, and thus predicted consequences for community dynamics depend on the effects of the new selection regime.

If species vary in dispersal ability, the dispersal phase of an organism's life cycle might also involve selection. For example, some species might more readily disperse to and therefore initially occupy disturbed sites, while other species, once established, might competitively exclude the good dispersers. With disturbance varying over space and time, this scenario effectively involves spatially and temporally variable selection, with dispersal-based life-history trade-offs representing the low-level processes underlying potential species coexistence (Levins and Culver 1971, Tilman 1994).

5.5. SPECIATION

The key processes involved in most theories falling under the traditional conceptual umbrella of community ecology (i.e., with a focus on relatively small spatial scales) can be understood as some combination of selection, drift, and dispersal. Thus, in a great many studies, speciation can be safely ignored. However, with a view of communities that extends to larger spatial and temporal scales (Elton 1927, Ricklefs and Schluter 1993a, Gaston and Blackburn 2000, Wiens and Donoghue 2004), speciation is required in a logically complete set of processes. Speciation is obviously a critically important ingredient in any process-based account of large-scale spatial variation in species diversity (Ricklefs and Schluter 1993a, Wiens and Donoghue 2004, Butlin et al. 2009, Wiens 2011, Rabosky 2013). Note that I do not consider extinction as a separate high-level process, because it is simply one possible result of selection and drift in a given community, the same way that loss of an allele is not considered a process but rather one possible result of selection and drift within populations.

An expansion of the spatial scales of interest is not the only reason speciation is necessary in a conceptual framework for community ecology. First, even at smaller scales (e.g., $< 10^4$ km^2), especially on isolated oceanic islands, in situ speciation can be an important source of new species added to the community (Losos and Schluter 2000, Gillespie 2004, Rosindell and Phillimore 2011). In such cases, speciation joins immigration (via dispersal) as a quantitatively important means by which diversity can increase. Second, the massive recent expansion of microbial community ecology has forced researchers to consider speciation as a process of central ecological importance, given the rapidity with which new phenotypically distinct lineages can arise in microbial communities (Hansen et al. 2007, Costello et al. 2012, Nemergut et al. 2013, Kassen 2014, Seabloom et al. 2015). Finally, as mentioned earlier, the causes of many local-scale community patterns (e.g., variation in species diversity along environmental gradients) often cannot be understood without reference to speciation (Ricklefs and Schluter 1993a, Gaston and Blackburn 2000, Wiens and Donoghue 2004).

Although speciation is the community-level analogue of mutation, the analogy is quite loose, at best. To a first approximation, it is reasonable to model a mutation as a random genetic change arising in the genome of one individual at a given moment in time. This is, in fact, exactly how Hubbell (2001) first modeled speciation in his ecological neutral theory: with some small probability ν (the Greek letter "nu"), the speciation rate, one individual organism instantly becomes the first and only individual of a new species. However, speciation may happen in many different ways, with important consequences for the initial population size, distribution, and traits of the resultant new species (Coyne and Orr 2004, McPeek 2007, Butlin et al. 2009, Nosil 2012). The rate and mode of speciation may also depend on properties of the community itself (Desjardins-Proulx and Gravel 2012, Rabosky 2013). The topic of speciation as a macroevolutionary process has been thoroughly reviewed elsewhere (Coyne and Orr 2004, Nosil 2012). Here we can focus on the consequences of speciation for the basic community-level patterns described in Chapter 2.

Speciation in a given community, on its own, should increase species richness. Speciation occurring independently in different communities (e.g., in the

Galápagos Islands vs. the South American mainland) will increase beta diversity between these communities. In terms of species composition, speciation effectively adds new elements to our vector of species abundances, **A**, or new rows to the species × site table (see Fig. 2.2). As in the case of immigration, subsequent effects on species composition may occur via altered selection, although these are not primary effects of speciation per se but secondary effects of an initial change to community composition. Such secondary effects of any individual speciation event might be quite minimal if new species are ecologically very similar to existing ones, or they might be very strong if new species are ecologically distinct (McPeek 2007, Butlin et al. 2009).

Most interestingly, in my opinion, speciation can contribute to the generation of relationships between species diversity and environmental conditions, even at a local scale in some cases. These relationships may occur to the extent that either the rate of speciation or the time that speciation has had to operate varies according to environmental conditions. For example, using a phylogenetic approach with salamanders, Kozak and Wiens (2012) found that greater speciation rates in the tropics contributed to positive diversity-temperature relationships, while Wiens et al. (2007) found that maximal salamander diversity was found at mid-elevations owing to an early date of initial colonization relative to low- or high-elevation habitats. These studies and others are addressed in greater detail in Chapter 10.

5.6. A NOTE ON ECO-EVOLUTIONARY DYNAMICS

Following publication of the initial sketch of the theory presented here (Vellend 2010), Marc Cadotte (University of Toronto, personal communication) suggested to me that speciation per se was only one way to generate new ecologically relevant phenotypes, with adaptive evolution and plasticity (within species) also altering phenotype distributions in ways that could influence community structure and dynamics. This is certainly true. One might articulate an *evolutionary theory of ecological communities* as follows: at the same time that ecological drift, selection, dispersal, and speciation dictate community dynamics, genetic drift, natural selection, gene flow, and mutation drive evolutionary change within species, potentially altering the "rules" of the community-level game, and vice versa. This reciprocal feedback, sometimes with reference only to single species, has been termed "eco-evolutionary dynamics" and is the subject of a large and rapidly growing body of literature (Fussmann et al. 2007, Urban et al. 2008, Pelletier et al. 2009, Schoener 2011, Norberg et al. 2012), including a recent synthetic monograph (Hendry, forthcoming).

Microevolutionary change is not part of the general theory of ecological communities for two reasons. First, it fits comfortably as a low-level process that— like many others—can cause the form and strength of selection in communities to vary over space over time. Many such processes are discussed in the next section of this chapter. For example, in the same way that nonlinear responses of species to resource levels can internally create fluctuating selection in a community (Armstrong and McGehee 1980, Huisman and Weissing 1999, 2001), so can adaptation via natural selection within species alter selection among species in the community (Levin 1972). Pimentel (1968) referred to this as the "genetic feedback mechanism" by which stability of each species' population might be maintained (see also Chitty 1957).

Second, even if one wished to accord a position of primacy to intraspecific evolution in community ecology (for which I see no justification), I would still argue that we first need to establish a foundation of high-level processes before formally building in additional complexities. Until we know what to expect in simplified situations, it will be hopeless to try to figure out what to expect in more complex situations (Bell 2008). Along with many other researchers, I have been a strong proponent of the need to consider intraspecific genetic variation and evolution to understand particular ecological outcomes (Vellend and Geber 2005, Vellend 2006, Hughes et al. 2008, Urban et al. 2008, Drummond and Vellend 2012, Norberg et al. 2012). However, while models including eco-evolutionary feedbacks are clearly needed for understanding and predicting the dynamics of some systems, it is not clear to me whether the many specific models integrating ecological and evolutionary dynamics will ultimately lead to a general eco-evolutionary theory, or whether they can more usefully be thought of as case-specific models drawing on two general bodies of theory, one ecological and one evolutionary.

In short, in the same way that the general theory of evolution by natural selection stands on its own without requiring explicit incorporation of the multitude of processes underlying natural selection (e.g., competition, abiotic stress, parasites) or mutation (replication errors, radiation, chemical mutagens), the theory of ecological communities stands on its own without requiring explicit incorporation of evolution within species or any other low-level process. Because the species is typically the fundamental categorization used to distinguish individual organisms in community ecology, we consider a high-level process in play only when evolution produces a new species.

5.7. CONSTITUTIVE THEORIES AND MODELS IN COMMUNITY ECOLOGY

The massive wave of ecological theorizing of the 1960s (see Chap. 3) was motivated, at least in part, by a perceived need for general principles that would help organize an "indigestible mass of field and laboratory observations" (May 1976). My own motivation in developing the conceptual framework just described derived from a strong sense that the theories themselves were now so many and varied as to be just as indigestible as the observations they were meant to make sense of. Thirty-five years ago, McIntosh (1980) articulated a similar sentiment: "If the models and theory which are presumably to provide the simplifying and explanatory framework for the indigestible mass of ecological observations are themselves unassimilable by the ecological corpus, ecology may have need of a purgative." The problem is especially acute from the point of view of students, most of whom can understand the essential results of empirical studies but exceedingly few of whom can make sense of the often-byzantine nature of theoretical papers in ecology.

A sample of theories and models that fall under the theory of ecological communities is presented in Table 5.1. The table has 24 entries; a more comprehensive version could have at least three to four times as many (Palmer 1994). However, all such theories and models can be understood with reference to just four processes: selection (in a few different forms), dispersal, drift, and speciation (see Table 5.1). The main route to theory proliferation in community ecology is the seemingly endless number of low-level processes and variables (competition, predation, stress, disturbance, productivity, etc.), and ways they can be manifested, that can be implicated as important causes of selection. Thus, one of the main ways of simplifying things is to recognize when different theories implicate the same high-level form of selection. This is essentially what Chesson (2000b) has done for models aiming to explain stable local species coexistence: any such model must, at some level, involve negative frequency-dependent fitness (HilleRisLambers et al. 2012).

In addition to helping simplify our view of theory in community ecology, I believe that recognition of four high-level processes can also help us think through some verbal arguments more clearly. Disturbance as a low-level process provides an illuminating example. Field ecologists see natural and anthropogenic disturbances all over the place: windthrow, fires, floods, dirt mounds created by burrowing animals, and so on (Pickett and White 1985). It is obvious that disturbance is a force of selection in communities: the species thriving in recently disturbed sites are usually different from those thriving at the same site either before or long after the disturbance event. But how might disturbance affect local species diversity?

Some much-cited arguments posit that disturbance helps maintain diversity by slowing competitive exclusion via decreased species densities (Connell 1978, Huston 1979). However, competitive exclusion occurs when selection favors one species over another, and just slowing down this process does not eliminate the inevitable outcome of reduced diversity (Roxburgh et al. 2004, Fox 2013). Reducing the abundance of all species equivalently does, however, influence the process of ecological drift. In a comparison of two communities with the same long-term average community size, drift will occur more rapidly in the one with greater temporal fluctuations (Adler and Drake 2008). The community with fluctuating size has a lower "effective community size"

TABLE 5.1. A Sample of Constitutive Theories, Models, or Frameworks That Fall under the Theory of Ecological Communities

Theory/Model	Description in Terms of High-Level Processes	Low-Level Processes Involved	References
Theories focused largely on selection in local communities			
The competitive exclusion principle	Constant selection	Two species competing for the same resource cannot coexist because one or the other will inevitably have at least a small advantage.	Gause (1934)
R* theory	Selection: constant or negatively frequency dependent	Competition for multiple limiting resources; R* is the lowest level of a resource at which a species can persist; for one resource, the species with the lowest R* wins; negative frequency-dependent selection emerges from trade-offs among species in R* and rates of uptake of different resources.	Tilman (1982)
Enemy-mediated coexistence	Selection: negatively frequency dependent	Enemies (e.g., predators, pathogens) have their greatest effects on the most abundant species, thereby favoring species when rare; a predator density, P*, can be defined in a similar way as R*.	Holt et al. (1994)
Janzen-Connell effects	Selection: negatively frequency dependent and spatially variable	Originally proposed to explain tropical forest diversity: species-specific enemies accumulate around adult trees, preventing local regeneration of that species; at a very small spatial scale, this gives an advantage to rare species. Conceptually related to enemy-mediated coexistence.	Connell (1970), Janzen (1970)
Temporal storage effect	Selection: temporally variable and negatively frequency dependent	Three criteria: (1) species respond differently to the environment; (2) per capita intraspecific competition is greatest when a species is most abundant, and interspecific competition is greatest when a species is rare; (3) species have some means of persistence through "bad" periods (e.g., a seed bank).	Chesson (2000b)

continued

TABLE 5.1. *Continued*

Theory/Model	Description in Terms of High-Level Processes	Low-Level Processes Involved	References
Relative nonlinearity of competition	Selection: temporally variable and negatively frequency dependent	Species show different non-linear fitness responses to resource levels; fluctuations in resource levels created by the species themselves cause selection to fluctuate temporally, potentially allowing coexistence.	Armstrong and McGehee (1980)
Genetic feedback	Selection: temporally variable and negatively frequency dependent	Selection in the community disfavoring a particular species creates strong natural selection within that species for traits that allow it to recover.	Pimentel (1968)
Priority effects	Selection: positively frequency dependent	Initial colonists of a given site inhibit or prevent establishment of other species; could be due to life stages (e.g., established adult plants prevent regeneration from seed) or positive feedbacks between the species and the environment (e.g., plant-soil feedbacks).	Origin of concept unclear; reviewed in Fukami (2010)
Intransitive competition	Selection: frequency dependent	Coexistence can be maintained because each species is both competitively superior and competitively inferior to some others, akin to a game of rock-paper-scissors.	Gilpin (1975)
Multiple stable equilibria	Selection: positively frequency dependent	With potentially complex details, the crux here is positive feedbacks such that beyond some threshold state (e.g., lots of algae or lots of coral in a reef) the more abundant type of species rises to dominance.	Reviewed in Scheffer (2009)
Succession theory	Dispersal, selection, and drift	Succession is an umbrella term synonymous with community dynamics and so can involve any and all low-level ecological processes; the term is most often applied to changes following disturbance.	Pickett et al. (1987)
Niche theory	Selection (all forms)	An umbrella term for all selection-based models of interacting species.	Chase and Leibold (2003)

TABLE 5.1. *Continued*

Theory/Model	Description in Terms of High-Level Processes	Low-Level Processes Involved	References
Theories focused on selection involving multiple communities or explicit spatial considerations			
Spatial storage effect	Selection: spatially variable and negatively frequency dependent	Same essential criteria as for the temporal storage effect, except the environment varies in space instead of across time.	Chesson (2000b)
Intermediate disturbance hypothesis	Selection: constant, spatially and temporally variable	High disturbance represents strong constant selection favoring few species; absence of disturbance favors only a few strong competitors; intermediate disturbance slows competition and creates temporally variable selection.	Grime (1973), Connell (1978)
The hump-shaped diversity-productivity hypothesis	Selection: constant, spatially variable	Stress and competitive exclusion reduce diversity at low and high productivity, respectively, while constraints on coexistence are relaxed at intermediate productivity. (Note that diversity-productivity relationships have generated many additional hypotheses).	Grime (1973)
Species-energy theory	Speciation, drift, selection	Starting from observed relationships between species diversity and energy (e.g., potential evapotranspiration), many low-level processes have been proposed as important, including speciation rate and community size (inversely related to drift).	Wright (1983), Currie (1991), Brown et al. (2004)
Colonization-competition tradeoffs	Dispersal and highly localized temporally variable selection	Species good at colonizing "empty" (e.g., disturbed) sites are easily displaced by good competitors, the latter of which are not good at colonizing empty sites. Disturbance creates temporally variable conditions within sites.	Levins and Culver (1971)
Metacommunities: mass effects	Dispersal and spatially variable selection	Continual dispersal from "source" populations maintains "sink" population of many species in unfavorable conditions.	Leibold et al. (2004)

continued

TABLE 5.1. *Continued*

Theory/Model	Description in Terms of High-Level Processes	Low-Level Processes Involved	References
Metacommunities: patch dynamics	Dispersal and selection	This is an umbrella term for models in which colonization-extinction dynamics are important; colonization-competition models (see above) are the best-known example.	Leibold et al. (2004)
Metacommunities: species sorting	Spatially variable selection	Different species are at an advantage under different environmental conditions.	Leibold et al. (2004)
Theories involving drift and/or speciation			
Stochastic niche theory	Selection and drift	Niche theory with the addition of demographic stochasticity (i.e., drift).	Tilman (2004)
The species pool hypothesis	Speciation, dispersal, and spatially variable selection	The number of species found locally under different environmental conditions is determined by the number of species adapted to such conditions in the regional species pool (which was created by speciation and dispersal); spatially variable selection drives composition-environment relationships but not diversity-environment relationships.	Taylor et al. (1990)
The theory of island biogeography	Drift and dispersal	Local species richness results from a balance between colonization (dispersal) and extinction (via drift). Diversity is least on islands that are small (more drift) and isolated (less dispersal).	MacArthur and Wilson (1967)
Neutral theory	Drift, dispersal, and speciation in a metacommunity	Local species diversity, relative abundance distributions, and beta diversity result from a balance between speciation, dispersal, and drift.	Hubbell (2001)

Note: This is only a sample. Models and theories in community ecology have been presented under so many different headings that it is nearly impossible to be comprehensive.

(Vellend 2004, Orrock and Watling 2010), which suggests that disturbance may actually reduce diversity. However, disturbance is also clearly a low-level process causing temporally variable selection, and if negative frequency dependence emerges, then coexistence and diversity can be promoted (Roxburgh et al. 2004, Fox 2013). While the thrust of this argument does not *require* reference to high-level processes, I think the reframing in terms of drift and selection clarifies how the essential features of disturbance—from the point of view of its effects on species diversity—relate to other models within the broader field of community ecology.

5.8. SO WHAT?

There are days when I think, "this is all so obvious, so what's the big deal?" and other days when I think, "this is so useful, why isn't community ecology already conceptually organized this way?" One lens through which to view these questions is, again, that of evolutionary biology.

The theory presented here projects the four-process perspective of the modern evolutionary synthesis onto community ecology (Fig. 5.4; see also Chap. 4). In evolutionary biology, this synthesis was a profound achievement given the fundamental uncertainty at the time (the 1930s and '40s) about the nature of inheritance and the processes underlying evolutionary change (Mayr 1982). However, some 80 years on, the core idea that variation is produced by mutation and migration, and subsequently molded by natural selection and drift,

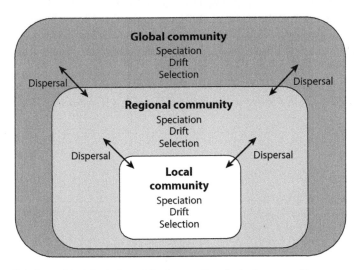

Figure 5.4. A graphical depiction of the theory of ecological communities across spatial scales.

is so well established as to seem quite obvious. Most contemporary studies in evolutionary biology are thus viewed not as *testing* the general theory but rather as *using* the general theory in the process of addressing more specific issues, such as the relative importance of different mechanisms of speciation (Nosil 2012) or the evolutionary consequences of anthropogenic disturbances (Stockwell et al. 2003).

The theory of ecological communities I have articulated is not such a profound achievement: it appears obvious right from the outset (at least to some). But perhaps it can be useful. By asking why the mutation-migration-drift-selection framework is still useful in contemporary evolutionary biology—despite being obvious given current knowledge—perhaps we can assess why a general theory of community ecology might be useful despite its obviousness. I posed just this question to a dozen or so evolutionary biologists, and three answers emerged repeatedly. First, the conceptual framework provides an invaluable pedagogical tool for understanding the essential features of evolution. Second, it provides a common conceptual framework for ensuring that researchers working on different specific evolutionary problems can understand one another and therefore avoid reinventing the wheel. Finally, the near-universal generality of the framework itself attracts young scientists to evolutionary biology (including some of those I spoke with).

I have already made an argument as to why I think the theory of ecological communities can help students make sense of what can seem like an overwhelming number and variety of constitutive theories and models (the first answer). Ecologists have often been criticized for frequently reinventing the wheel via the application of new terms to old ideas (Lawton 1991, Graham and Dayton 2002, Belovsky et al. 2004), so perhaps the second answer is also relevant to the theory of ecological communities (Tucker and Cadotte 2013). With respect to the third answer, one can hope only that a general theory may steer bright young minds to ecology.

CHAPTER 6

Simulating Dynamics in
Ecological Communities

So far I have treated theoretical ideas in community ecology in a largely qualitative manner. While many of the greatest conceptual advances in biology have
come via qualitative reasoning (Wilson 2013), mathematical models are essential to further progress for several reasons (Otto and Day 2011, Marquet et al.
2014). First, they provide critical tests of our intuition: if we are unable to produce a quantitative representation of a particular phenomenon, we must wonder
whether there are holes in our qualitative logic. They also force us to be explicit
about simplifying assumptions, which we might not even realize we are making.
Models provide a means of exploring the sensitivity of particular outcomes (e.g.,
stable coexistence) to various kinds of conditions (e.g., the pattern of temporal
environmental fluctuations) and of generating quantitative predictions or fore-
casts for what we might expect to see in nature. Finally, they provide a vehicle
for conceptual unification in that we can hope to show that many specific models are special cases of one or a few more general models.

My aims in this chapter are (i) to illustrate expected community dynamics
and patterns under different combinations of high-level processes—that is, to
provide the theoretical basis for the hypotheses and predictions presented in
Chapters 8–10; (ii) to support the argument that the core elements of most
models in community ecology are just different combinations of a few basic
ingredients; and (iii) to provide readers with the tools to explore scenarios on
their own—a very effective means for learning about the process-pattern link—
without requiring much in the way of mathematical ability.

This chapter is presented in a decidedly nonstandard way. In addition to presenting theoretical predictions (as in many standard texts), I also present and

explain the raw computer code, allowing readers to generate and explore these predictions on their own (not standard). Readers not interested in the latter can skip ahead to the figures, which present the main model predictions.

6.1. GETTING STARTED WITH MODELING

Many modeling frameworks can be applied to predicting and understanding the dynamics and structure of ecological communities. A distinction is commonly made between analytical models and numerical or simulation models. (The analytical vs. numerical distinction often refers to how one works with an equation or set of equations rather than how they are initially constructed, but I refer to them as different kinds of models for simplicity.) Analytical models are those with a "closed-form" or "general" solution, meaning that once equations are written, they can be "solved" to predict the outcome(s) of interest (e.g., population size of a given species) at any arbitrary time in the future based on initial conditions and the values of the model parameters. The models for exponential population growth, logistic growth, and the Lotka-Volterra equations presented in Chapter 3 fall into this category. A great advantage of such models is that their behavior can be understood quite precisely, with outcomes traceable to specific causes. They can also be applied with just paper and pencil, and, perhaps most important, they are aesthetically very pleasing to a mathematician.

Beyond a certain degree of complexity, it becomes nearly impossible to find (or make sense of) closed-form solutions to analytical models, in which case researchers use simulations. In a simulation model, a researcher first defines the initial state of the system (e.g., the population size of each interacting species), as well as a set of equations or rules that govern subsequent changes, and then uses a computer to explore dynamics under different scenarios. In addition to allowing incorporation of greater complexity, simulation models also have the appeal of permitting a more concrete mental mapping of what happens in the model to what happens in nature. For example, in lines of simulation code I can "see," in my mind's eye, a canopy tree dying, seeds being dispersed, and then saplings competing to replace the fallen canopy tree (Pacala et al. 1993). For someone adept at mathematics, some intuitive scenarios like these can be represented in an analytical model, provided the scenarios are not overly complex (Otto and Day 2011). But most ecologists are not sufficiently proficient at mathematics to convert an ecological scenario into equations, or even to easily follow experts through the analysis of such models.

For the reasons just described, I believe that simulations of even simple ecological models (i.e., for which analytical versions are possible) can be of great pedagogical utility in allowing *all* ecologists to explore theoretical community dynamics on their own, which is far more conducive to learning than simply following an expert walking through a model. After learning how to convert "pseudocode" (i.e., simple, structured verbal instructions for how a simulation should work) into functional computer code, a world of theoretical possibilities is opened up for exploration. Learning how to do this requires a nontrivial initial effort—but no more than mastering the dozens of bells and whistles on a new smartphone (no one seems to have trouble finding the time for that)—and it requires very little mathematics. To be clear, this is not an argument that sophisticated mathematics is not of immense importance in ecology, but it is an acceptance of the fact that many practitioners of ecology find the theoretical literature inaccessible, and probably always will. I hope that by exploring ecological simulations readers can gain easier access to the core theoretical ideas in community ecology. Perhaps some will even be inspired acquire the skills needed to build analytical models of their own.

All the simulations presented here were conducted using the freely available R programming language (R Core Team 2012). To maximize accessibility, I assume no knowledge of R whatsoever and provide detailed explanations for the simulation code. Code for the first model (neutral dynamics in one local community) is provided here (see Box 6.1) so that readers can get a sense of how things work without needing to refer elsewhere. All other code (Online Boxes 1–8) can be accessed at http://press.princeton.edu/titles/10914.html. To reproduce particular figure panels, readers will need to set parameters to particular values before running simulations.

6.2. LOCAL COMMUNITY DYNAMICS: ECOLOGICAL DRIFT

I begin with as simple a scenario as I can imagine, using a model known as the *Moran model*, originally designed to understand changes in allele frequencies in populations (Moran 1958, see also Hubbell 2001, Nowak 2006). I use this model as the foundation for all the simulations in this chapter to demonstrate that the essence of each of a very large number of models can be captured by making just one or a few fairly minor adjustments to a core model. Mathematically savvy readers should be forewarned that while all models are presented here in a way that captures their key ecological features, in some cases (e.g., colonization-competition tradeoffs, island biogeography) my implementations do not follow the mathematical conventions used in their original formulations.

For dynamics in a neutral, closed community with no speciation, the following is the recipe, or pseudocode, for the Moran model:

    1. Specify an initial community of J individuals divided among S species; each species i has N_i individuals.

    2. Select an individual at random to die.

3. Select an individual at random to produce one offspring, which replaces the dead individual.

4. Repeat from step 1.

The beauty of this model lies in its simplicity: a wide range of dynamics and patterns can be produced just by changing the rules at step 3 for selecting an individual to reproduce. Modifying step 2 could achieve the same goal. On the surface, step 3 might seem a bit odd biologically (one offspring from one individual), but it is equivalent to assuming that all individuals produce many offspring, only one of which will become a new recruit to the community at a given point in time. Because we will be working with simulations, I will henceforth use the notation and font (`Courier`) that appears in the R code to refer to variables and parameters in the code itself. Variables and parameters that don't appear explicitly in the R code will remain in the text font.

The state of a modeled community is specified by the vector of species abundances, $[N_1, N_2 \ldots N_S]$, in which N_i is the abundance or population size of the ith species. With an assumption of constant J, we can keep track of species frequencies as described in Chapter 5. With only two species, `freq.1` $= N_1/J$, and `freq.2 = 1 − freq.1`, so we need to keep track of only one of the species frequencies as a complete specification of what's happening in the community. That is, pattern in the community is fully accounted for by a single "response" variable. We can consider J repetitions of this cycle as one time step, which for simplicity we can think of as a "year" for organisms with an average life span of one year.

Let's start with the neutral model just described in a community with two species, 1 and 2. R code for this model is shown in Box 6.1, which is deliberately a bit longer than necessary to permit the addition of selection in as simple a way as possible. (To use R to explore analytical models like those discussed in Chap. 3, see Stevens (2009).) Since all individuals have the same probability of being chosen to reproduce, the probability that the reproducing individual will be of species 1 is simply the frequency of this species in the community: `Pr.1` = `freq.1` $= N_1/J$. The resulting dynamics are due solely to drift. Species frequencies bounce up and down at random until one or the other species goes extinct (Fig. 6.1). Drift is slower in larger communities, and the probability that a given species eventually "wins" is equal to its initial frequency (Fig. 6.1; Kimura 1962, Hubbell 2001).

6.3. LOCAL COMMUNITY DYNAMICS: SELECTION

Selection occurs when fitness differs between species, that is, when a species' probability of being chosen to reproduce is different from its frequency: `Pr.1` $\neq N_1/J$. Imagine a community with 10 individuals of species 1 and 30 of spe-

BOX 6.1.

SIMULATING NEUTRAL DYNAMICS IN A LOCAL
TWO-SPECIES COMMUNITY (WITH NO SPECIATION) USING R

R code is presented in Figure B.6.1, with explanations for each line of code (referenced by numbers on the left) provided here:

```
1.    J <- 50
2.    init.1 <- J / 2
3.    COM <- vector(length = J)                      (1) Specify initial community
4.    COM[1:init.1] <- 1; COM[(init.1 + 1):J] <- 2       (& time instructions)
5.    num.years <- 50
6.    year <- 2

7.    freq.1.vec <- vector(length = num.years)       Set up vector for
8.    freq.1.vec[1] <- init.1 / J                     data collection

9.    for(i in 1:(J * (num.years - 1))) {

10.       freq.1 <- sum(COM == 1) / J
11.       Pr.1 <- freq.1
12.       COM[ceiling(J * runif(1))] <- sample(c(1, 2), 1,
             prob = c(Pr.1, 1 - Pr.1))               (2,3,4)
                                                      Run simulation
13.       if (i %% J == 0){
14.          freq.1.vec[year] <- sum(COM == 1) / J
15.          year <- year + 1
16.       }
17.    }

18.    plot(1:num.years, freq.1.vec, type = "l", xlab = "Time",   Create a graph
19.    ylab = "Frequency of species 1", ylim = c(0, 1))            of simulation
```

Figure B.6.1. R code to simulate neutral dynamics in a local two-species community with no speciation. The numbers on the right correspond to the pseudocode (see the text), and the numbers on the left serve as references for the explanations in Box 6.1. To function in R, these numbers must be removed. Fully annotated code that can be used in R is provided in Online Box 1 at http://press.princeton.edu/titles/XXXX.html.

1. Define the local community size, J. J is defined as an object, and <- places the number 50 in this object.
2. Define the initial population size of species 1, init.1. By default, the initial population size of species 2 will be $J - init.1$.
3. Create an empty vector of length J to represent the community, and call it COM.
4. Assign individuals 1 through init.1 in COM to be species 1 and the remaining individuals to be species 2.
5. Set the number of years over which to run the simulation.
6. Define the first year to be simulated as year 2 (the initial specified community

(Box 6.1 continued)

will be year 1). If we want to record output each year, as opposed to after each individual birth-death event, we need this to keep track of years in the loop that starts on line 9.

7. Create an empty vector to hold the output. We need to keep track only of the frequency of species 1, so we call this vector `freq.1.vec` (the frequency of species 2 is 1 – the frequency of species 1).

8. Record the initial frequency of species 1 in the first element of `freq.1.vec`.

9. Initiate the simulation. Since each "year" involves J birth-death events, and since the first year is already specified by the initial conditions, we need to go through the loop (i.e., repeat the birth-death cycle) $J*(num.years-1)$ times to simulate the specified number of years. The variable `i` keeps track of how many times we've gone through the loop: the first time through the loop, `i = 1`; the second time, `i = 2`, and so on.

10. Calculate the current frequency of species 1 (`freq.1`). `COM == 1` creates a vector with a "TRUE" (read quantitatively as 1) for any element equal to 1, and a "FALSE" (quantitatively zero) otherwise (in this case when it is a 2). So, taking the sum of `COM == 1` gives us the current population size of species 1, and dividing by J gives the frequency.

11. `Pr.1` is the probability that an individual of species 1 is chosen to reproduce; since this model is neutral, it is equal to `freq.1`. (Models with selection will use a different equation.)

12. Select an individual to die and replace it with an individual of the species chosen to reproduce. `runif(1)` draws one random number from a uniform distribution between 0 and 1, so `J*runif(1)` generates a random number between 0 and J. But we need an integer to select an individual from the community, and the `ceiling` function rounds up our random number to provide a random integer between 1 and J. This is the individual that will die. On the right-hand side, we determine the species identity of the reproducing individual based on `Pr.1`. `c(1,2)` concatenates the numbers 1 and 2 together in a vector, and we sample 1 number from this vector based on the probabilities `Pr.1` for species 1 and $1 - Pr.1$ for species 2. Thus, we choose a 1 or a 2 to replace the dead individual.

13–15. After each sequence of J deaths, record data. `i %% J` returns the remainder of `i` divided by J, and each time that J deaths have occurred this will be equal to zero, so this is an efficient way to tell R to "stop" the program, record the frequency of species 1 (line 14), and increment the year tracker by 1 (line 15).

16–17. Terminate the `if` loop and the `for` loop.

18–19. Plot the results. `1:num.years` are the data for the *x*-axis, and `freq.1.vec` the data for the *y*-axis. `type="l"` specifies a line graph, `xlab` and `ylab` specify the axis labels, and `ylim` specifies limits on the *y*-axis values.

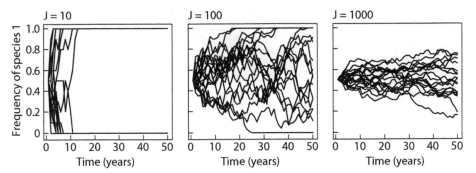

Figure 6.1. Dynamics of two-species communities under pure ecological drift. Each panel shows 20 independent simulations using a given community size (J) and a starting frequency of species 1 (init.1) of 0.5 (see code in Fig. B.6.1 and Online Boxes 1 and 2). Under pure ecological drift, species frequencies fluctuate randomly until one species dominates. Drift is stronger in smaller communities, and the probability that a species will ultimately dominate is equal to its initial frequency.

cies 2. Then, `freq.1` = 10/40 = 0.25 and `freq.2` = 30/40 = 0.75. If the per capita reproductive outputs (i.e., fitnesses) of the two species over a given time step are 20 and 10, respectively, we expect species 1 to produce $10 \times 20 = 200$ offspring and species 2 to produce $30 \times 10 = 300$ offspring. If we are going to choose just one of these offspring at random to replace a dead individual, the probability that this offspring will be of species 1 is equal to the proportion of all offspring that were produced by species 1: $(10 \times 20)/(10 \times 20 + 30 \times 10) =$ $200/(200 + 300) = 0.4$. So, because species 1 has higher fitness, its probability of producing a successful recruit in the community (`Pr.1` = 0.4) is greater than its frequency (`freq.1` = 0.25). To generalize, if the fitnesses of species 1 and 2 are `fit.1` and `fit.2`, respectively, then `Pr.1` = `fit.1*freq.1/` `(fit.1*freq.1 + fit.2*freq.2)` (Ewens 2004). By dividing the numerator and denominator of this equation by `fit.2` we can see that only the ratio of the two fitness values matters, not their absolute values:

$$Pr.1 = (fit.1/fit.2)*freq.1/((fit.1/fit.2)$$
$$*freq.1 + freq.2)$$

We can call `fit.1/fit.2` the fitness ratio (`fit.ratio`). The R code for local models with selection (Online Box 2) includes only `fit.ratio` and not `fit.1` and `fit.2` as separate parameters.

In the following sections on local community dynamics with selection, we will explore situations in which the relative fitnesses of the two species (`fit. ratio`) depends on two parameters: (1) the fitness ratio averaged across all possible species frequencies (`fit.ratio.avg`), which quantifies the strength of constant selection, and (2) the direction and strength of the relation-

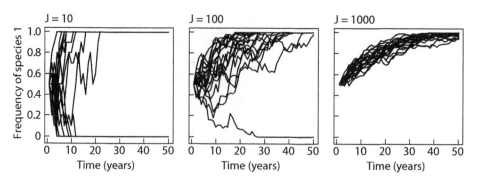

Figure 6.2. Dynamics of two-species communities under constant selection, favoring competitive exclusion of species 2 by species 1 (`fit.ratio.avg` = 1.1, `freq.dep` = 0). Each panel shows 20 independent simulations using a given community size (J) using the R code in Online Box 2. Selection favors dominance by species 1, but this is not certain at small community size. See the text for parameter definitions.

ship between fitness and species frequencies (`freq.dep`), which quantifies frequency-dependent selection. The R code for these models is presented in Online Box 2. If `fit.ratio.avg` = 1 and `freq.dep` = 0, the model is entirely neutral and so equivalent to the code in Box 6.1. In exploring scenarios with selection, we will first assume that `freq.dep` = 0, with the strength of constant selection set by `fit.ratio.avg`. We will then explore various scenarios of frequency-dependent and temporally variable selection. By exploring simulations with different values of J, we also can stay ever-cognizant of the fact that when a community has relatively few individuals (i.e., J is small), the outcomes expected based on fitness differences among species are not guaranteed. That is, stochastic drift can, in principle, overwhelm deterministic processes.

6.3.1. Competitive Exclusion via Constant Selection

If species 1 has consistently greater fitness than species 2 (i.e., `fit.ratio` > 1 always), species 1 will tend to competitively exclude species 2, and vice versa (Fig. 6.2). To simulate this situation, the only changes required to the R code in Box 6.1 are the assignment of the average fitness ratio and alteration of the equation for `Pr.1` (i.e., step 3 in the pseudocode). For example, keeping `freq.dep` = 0, we can set `fit.ratio.avg` = 1.1, giving species 1 a small fitness advantage (see Online Box 2).

6.3.2. Stable Coexistence via Negative Frequency-Dependent Selection

If the fitness of species 1 is greater than the fitness of species 2 when species 1 is rare, and vice versa, each species effectively has a relative advantage when rare, and there should thus be a stable equilibrium point at which both species

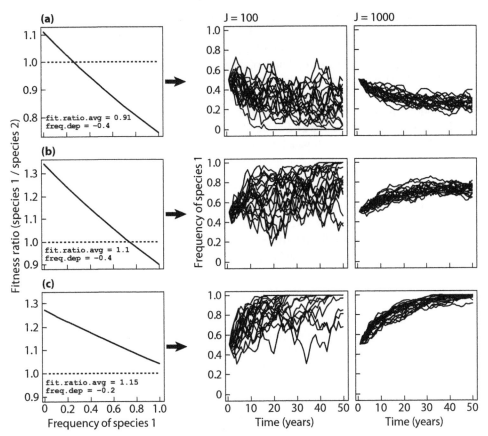

Figure 6.3. Dynamics of two-species communities under negative frequency-dependent selection. Each panel in the two columns on the right shows 20 independent simulations using a given community size (J). In the first two rows (a, b), frequency-dependent selection is strong enough relative to average fitness differences to lead to a stable equilibrium with both species present (the intersection of the solid and dotted lines in the left panels) and thus species coexistence. In the third row (c), coexistence is not possible despite negative frequency-dependent selection because species 1 always has an advantage. See the text for parameter definitions, and Online Box 2 for the R code.

have nonzero abundance (Chesson 2000b). That is, the two species should coexist via negative frequency-dependent selection. For stable coexistence, it is not enough for species' fitnesses to be negatively related to their frequencies, but this tendency must be strong enough relative to any average fitness difference such that `fit.ratio > 1` when species 1 is extremely rare, and vice versa (Adler et al. 2007; Fig. 6.3). In the left panels of Fig. 6.3, the slope of the line quantifies the strength of negative frequency-dependent selection (Chesson's "niche differences"), and the average position of the line on the *y*-axis quantifies the strength of constant selection (Chesson's "fitness differences"). These two properties of the relationship between fitness and frequency summarize, in a nutshell, the crux of "modern coexistence theory" (HilleRisLambers et al. 2012). These scenarios can be implemented using the R code in Online Box 2.

To simulate negative frequency-dependent selection, we need to specify the relationship between fitness and frequency. Most simply, the ratio of fitnesses of species 1 and 2 (`fit.ratio`) should be related negatively to the frequency of species 1 (Fig. 6.3, left panels). The details are not critical to understanding the key outcomes of this scenario, but to ensure symmetry in the relative (dis)advantages experienced by the two species, we can specify the logarithm of `fit.ratio` as a linear function of species' frequencies (see App. 6.1 and Online Box 2). The parameter `freq.dep` specifies the slope of this relationship. The parameter `fit.ratio.avg` is the value of `fit.ratio` when species frequencies are both 0.5, effectively shifting the fitness-frequency relationship up or down on the *y*-axis in the left panels of Fig. 6.3.

6.3.3. Temporally Fluctuating Selection

In the simple simulations of negative frequency-dependent selection so far with large communities, community composition (i.e., species frequencies) converges smoothly toward a stable equilibrium. However, this general "rule" about stable coexistence emerging from negative frequency dependence applies to *long-term* expected advantages when species are rare, which may arise via mechanisms that involve temporal fluctuations. Some important specific conditions need to be met in particular models (Chesson 2000b, Fox 2013), but most simply, fitness (dis)advantages can fluctuate over time such that each species spends enough time at an advantage to overcome population declines experienced while at a disadvantage. An additional condition is "buffered" population dynamics: the presence of some means by which a species can "hang on" (e.g., dormant propagules) when conditions are strongly unfavorable (Chesson 2000b) and/or some mechanism to prevent a dominant species from taking over completely when favored (Yi and Dean 2013).

Fitness variation can be driven by extrinsic fluctuations (e.g., climate) or by environmental fluctuations driven intrinsically by the organisms themselves (e.g., via resource use; Armstrong and McGehee 1980, Huisman and Weissing 1999, 2001). Just for purposes of illustration (i.e., without getting into the nitty-gritty mathematical requirements for indefinite coexistence), we can see that giving each of the two species a fitness advantage in alternating 10-year periods (e.g., via climate fluctuations) can contribute to the possibility that each species will tend to bounce back from being rare (Fig. 6.4; see Online Box 3 for R code). A comparable situation involving environmental differences across space is explored in subsequent sections.

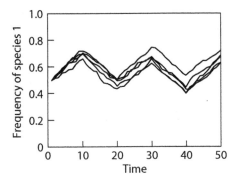

Figure 6.4. Dynamics of two-species communities under temporally fluctuating selection. Five independent simulations are shown with $J = 4000$ and alternating fitness ratios of 1.1 and 0.91 (the reciprocal of 1.1) each 10 years (R code in Online Box 3). In these simulations, eventually one or the other species will "win" (via drift); long-term stable coexistence depends on additional criteria, such as the ability to buffer population losses.

6.3.4. The Potential for Cyclical Dynamics

Many ecological models also have equilibria that are not represented by single points (e.g., a given species frequency) but, rather, by regular fluctuations or cycles that are expected to continue indefinitely. Most such models involve an explicit accounting of trophic interactions (e.g., predator-prey cycles), but the key feature is a tendency to repeatedly "overshoot" somewhat of an equilibrium point, and this propensity emerges even in some single-species population models (May 1974) or in multispecies models with only competition (Gilpin 1975). There are exceedingly few empirical examples of limit cycles caused only by "horizontal" interactions among species, so the simulations here are presented largely for theoretical interest, capturing an important class of possible community dynamics (i.e., this topic does not feature prominently in the empirical Chaps. 8–10).

Using our simulation framework, we can illustrate "overshoot" dynamics by adding a delay in the response of species' fitnesses to changes in their frequencies. Specifically, if we use the species' frequencies at the start of a "year" (i.e., a sequence of J deaths) to calculate a `fit.ratio` that will stay constant for the whole year, rather than always being adjusted according to current species frequencies, very strong frequency dependence can create indefinite fluctuations (Fig. 6.5). To implement this simulation in R (see Online Box 4), instead of one time loop with `J*num.years` steps, we create two nested loops, the first with `num.years` steps and the second with `J` steps, and `fit.ratio` defined between the start of the first and the second loops. Cycles can also emerge in models with "intransitive competition" among three species (Gilpin

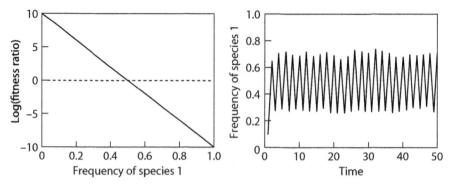

Figure 6.5. Dynamics of a two-species community under very strong "delayed" frequency-dependent selection. In this simulation (Online Box 4) species fitnesses stay constant for a year at a time (i.e., they are not updated after every death), and so species frequencies repeatedly overshoot the quasi-equilibrium of 0.5. Fluctuations are evident only with `freq.dep` < −10. One simulation is shown here with J = 500, `freq.dep` = −20, and initial frequency = 0.1. Note that the y-axis in the left panel is shown with a log scale, so the reference line for equal fitness across species is at zero instead of 1. See the text for parameter definitions.

1975), which means that the three species effectively play a game of rock-paper-scissors, which is inherently frequency dependent: rock beats scissors, paper beats rock, and scissors beats paper. As any one of the three species (e.g., rock) increases in abundance, a different species (e.g., paper) gains an advantage, and the three species take turns dominating the community (Gilpin 1975, Sinervo and Lively 1996, Vellend and Litrico 2008).

6.3.5. Priority Effects and Multiple Stable Equilibria via Positive Feedbacks

If species' fitnesses are positively related to their frequencies, such that `fit.ratio` < 1 when species 1 is rare, and vice versa, then whichever species starts at a high frequency will tend to exclude the other (i.e., there are strong priority effects). All that is needed to simulate priority effects is to switch the sign of `freq.dep` (in Online Box 2) to be positive (Fig. 6.6). This model illustrates one of the simplest possible situations in which one can generate multiple stable equilibria—in this case, dominance by one species or the other. Many more complex and often system-specific models also predict multiple stable equilibria, often with respect to multiple biotic and abiotic attributes of an eco-system simultaneously, such as dominance versus near-absence of particular functional forms (e.g., aquatic macrophytes) and major shifts in environmental variables (e.g., water clarity) (Scheffer 2009). At the core of all such models are positive feedbacks of one sort or another (e.g., facilitation of some species

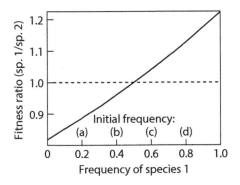

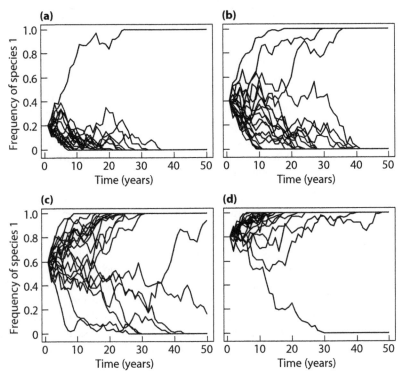

Figure 6.6. Dynamics of two-species communities under positive frequency-dependent selection. This is a simple example of multiple stable states: one or the other species is deterministically predicted to dominate, depending on their initial frequencies. At initial frequencies of species 1 > 0.5, species 1 has a fitness advantage, and vice versa (`freq. dep = 0.4`, `fit.ratio.avg = 1`). In all cases, J = 100, and each panel shows 20 independent simulations using different initial frequencies: (a) 0.2, (b) 0.4, (c) 0.6, and (d) 0.8. In (b) and (c), initial frequency is sufficiently close to 0.5 that frequencies can often drift to one side or the other of 0.5, thus favoring competitive exclusion of one species or the other. In (a) and (d), initial frequencies are farther from 0.5, thus leading to more predictable competitive exclusion. Panels (a)–(d) were produced using the R code in Online Box 2. See the text for parameter definitions.

by others), so the basic lesson here is more general than the extremely simple model might first appear.

6.4. LOCAL COMMUNITIES LINKED BY DISPERSAL

At the end of this chapter (Sec. 6.5), we will simulate dynamics in communities with larger numbers of species under the influence of dispersal and speciation. However, some of the key community-level consequences of dispersal, as well as the consequences of spatially variable selection, can most easily be understood by continuing with simulations of the dynamics of two species now in two or more local communities, or habitat "patches." Online Box 5 shows the R code needed to simulate the dynamics of two species in an arbitrary number of habitat patches (`num.patch`), each with J individuals. The entire set of patches is the metacommunity. Within patches, local selection (of any form) operates exactly as described previously, but the patches are now potentially linked by dispersal. Specifically, with probability m (the dispersal parameter), the reproducing individual at step 3 of the Moran model (described at the start of the chapter) is chosen at random from the entire metacommunity (i.e., the entire set of patches considered as a single unit), rather than from the local patch where a mortality event occurred.

6.4.1. The Interaction of Drift and Dispersal

In a purely neutral model with no dispersal (Fig. 6.7a), composition in each community drifts randomly, creating compositional variation (i.e., beta diversity) among patches. With dispersal, the dynamics are no longer independent among patches. Composition (in this case just the frequency of species 1) in the entire metacommunity is still prone to drift, but given the large size of the metacommunity (`num.patch*J` = 10*J in Fig. 6.7), drift at this level is relatively slow. With very high dispersal (Fig. 6.7c), there is essentially no biological meaning to "local" communities, except that they represent a subset of one larger community that behaves as a single entity. At intermediate levels of dispersal (Fig. 6.7b), composition in each patch can fluctuate about the average composition of the metacommunity, which is itself prone to drift.

6.4.2. The Interaction of Dispersal and Selection

With two habitat patches, each selectively favoring one of the two species, both species can potentially coexist indefinitely in both patches if there is some dispersal. In other words, spatially variable selection can promote diversity both within and across patches. If selective (dis)advantages are symmetric across patches—i.e., species 1 has the same selective advantage in patch 1 as species

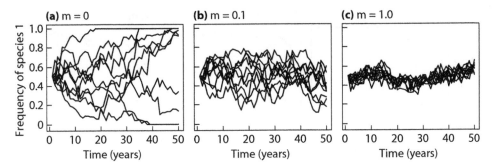

Figure 6.7. Dynamics of two species in 10 local communities (habitat patches) in a purely neutral model with $J = 100$ in each local community, and different levels of dispersal (m). See Online Box 5 for the R code. With no dispersal ($m = 0$), drift creates high beta diversity among patches (a); higher levels of dispersal (b, c) homogenize composition across patches.

2 does in patch 2—species coexistence across the metacommunity is expected regardless of the level of dispersal. With no dispersal (Fig. 6.8a), each species excludes the other in one of the patches. With increasing levels of dispersal (Fig. 6.8b,c), the selectively disadvantaged species in a given patch occurs there at increasingly higher abundance because of constant influx from the patch where it has an advantage.

When selection is asymmetric among habitat patches, such that species 1 has a greater selective advantage in patch 1 (e.g., fitness ratio = 1.5) than species 2 has in patch 2 (e.g., fitness ratio = 1.1^{-1}), high enough dispersal can lead to the extinction of species 2. In this scenario, in the absence of dispersal (Fig. 6.9a),

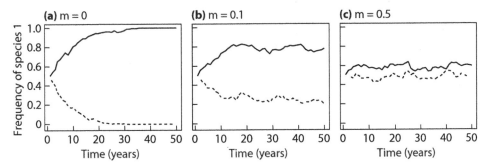

Figure 6.8. Community dynamics in two habitat patches, one selectively favoring species 1, the other favoring species 2. The fitness ratio is 1.2 in one patch (solid line) and 1.2^{-1} in the other (dashed line). $J = 1000$ in each community. See Online Box 5 for the R code. Spatially variable selection promotes coexistence of the two species, with dispersal ($m > 0$) countering local selection and homogenizing composition across patches.

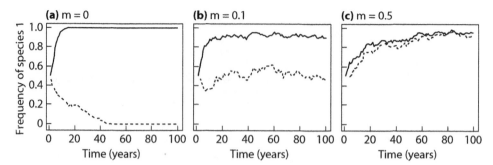

Figure 6.9. Community dynamics in two habitat patches, one selectively favoring species 1, the other favoring species 2. In this case, fitness ratios are asymmetric: 1.5 in one patch (solid line) and 1.1^{-1} in the other (dashed line). $J = 1000$ in each community. See Online Box 5 for the R code. Coexistence in the metacommunity (the two patches collectively) is maintained at relatively low dispersal (a, b), but very high dispersal (c) allows species 1 to dominate the entire metacommunity.

each species excludes the other in the patch where it has a selective advantage. With increasing dispersal (Fig. 6.9b,c), individuals of species 2 recruiting to patch 1 find themselves at a severe selective disadvantage, while recruits of species 1 in patch 2 are at a less severe disadvantage. Consequently, beyond a certain level of dispersal, the composition of patch 2 is "pulled" toward that of patch 1, ultimately leading to the demise of species 2 across the whole metacommunity (Fig. 6.9).

6.4.3. Selection at the Dispersal Stage: Competition-Colonization Trade-Offs

The models described in Sections 6.4.1 and 6.4.2 (code in Online Box 5) implicitly assume that all species have the same dispersal ability. When new recruits are drawn from the metacommunity rather than the local community, the draw is random. However, just as different species might make different per capita contributions to the local offspring pool—thus having different probabilities of local success—species might also vary in their relative contributions to the pool of offspring that disperses among patches. This situation can be modeled by effectively considering dispersal ability as a component of fitness, specifically with respect to defining the probability that a new recruit drawn from the whole metacommunity is of one species or the other.

A new "dispersal" fitness ratio (`fit.ratio.m`) can be defined (see Online Box 6), and then when dispersal is implemented in the `for` loop, we first calculate the frequency of species 1 in the entire metacommunity and then use that value in conjunction with `fit.ratio.m` to calculate `Pr.1`, exactly as done when recruitment happens locally. This process introduces the possibility

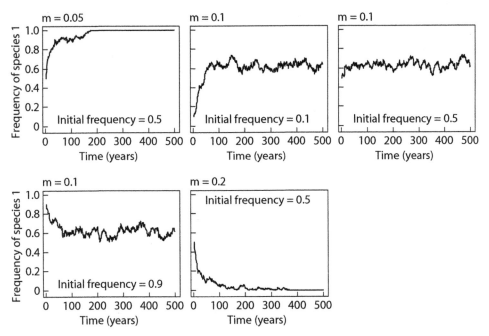

Figure 6.10. Dynamics in a metacommunity of two habitat patches ($J = 1000$ per patch), both of which selectively favor species 1 (`fit.ratio.avg` = 1.2). There is dispersal between patches ($m > 0$), and at the dispersal stage species 2 has a "fitness" advantage (`fit.ratio.m` = 1/5) (see Online Box 6). Because local parameters are identical in the two patches, only the metacommunity-level frequency is shown. Different initial frequencies are shown for $m = 0.1$ to demonstrate the tendency for convergence toward an equilibrium frequency of ~0.6. When dispersal is low ($m = 0.05$) species 1 "wins"; when dispersal is high ($m = 0.2$) species 2 wins. These simulations capture the key feature of competition-colonization trade-off models, with species 1 better at "competition" (`fit.ratio.avg` > 1) and species 2 better at "colonization" (`fit.ratio.m` < 1).

of counteracting fitness components: species 1 might have an advantage locally (`fit.ratio` > 1 in both patches), while species 2 might have a superior dispersal ability (`fit.ratio.m` < 1). Fitness components along these lines have often been referred to as "competition ability" and "colonization ability," respectively. With a sufficiently strong trade-off between these "abilities," and neither too little nor too much dispersal, species coexistence in the metacommunity is possible (Levins and Culver 1971, Tilman 1994). In mathematical terms, with identical local selection in multiple patches, the metacommunity is like one big patch in which some recruitment events ("local" ones) favor species 1, and others ("dispersal") favor species 2. Nonetheless, the results illustrate the potential for life-history trade-offs to promote the maintenance of diversity (Fig. 6.10).

6.4.4. Overview of Models with Dispersal

These fairly simple simulations, each version of which involves just small modifications to a core model, illustrate many well-known and seemingly disparate
theoretical results. First, spatially variable selection (i.e., via environmental
heterogeneity) can be a potent force in maintaining diversity (Levene 1953).
Second, dispersal can often maintain sink populations of species, thereby el-
evating diversity at the local scale (MacArthur and Wilson 1967). Dispersal
also causes community composition across habitat patches to converge (i.e.,
beta diversity will decline), regardless of local selection (Hubbell 2001, Chave
et al. 2002). If there is some asymmetry in the selective advantages enjoyed
by different species in different places, very high levels of dispersal can ultimately erode diversity across the metacommunity (Mouquet and Loreau 2003).
Finally, if dispersal ability and local selective advantages are negatively correlated across species (i.e., there is a trade-off), species coexistence can be
maintained despite spatially homogenous local selection (Levins and Culver
1971, Tilman 1994).

6.5. MODELS WITH SPECIATION

Two final points can be illustrated by incorporating speciation into simulations
of community dynamics. First, a higher speciation rate increases species richness, as well as the evenness of species abundances. Second, local-scale diversity is greater when immigrants are drawn from a regional species pool where
the speciation rate has been higher. Here we will first simulate "regional" community dynamics as a balance between speciation and drift. Then, using different regional communities as sources of immigrants (i.e., "mainlands"), we
will simulate a local balance between immigration (via dispersal) and drift, as
in the classic island biogeography model (MacArthur and Wilson 1967). In neither case will there be any selection so as to focus on the effects of speciation
without additional complications.

A neutral model with speciation looks much like the "local" neutral model
already encountered (see Box 6.1), except that with a small probability nu
(often denoted by the Greek symbol v) a new recruit to the community will be
of a new species (see Online Box 7). We need to keep track of different species
using different numbers, and the COM vector may have many different species
in it. Using this code, we can see that an increasing speciation rate leads to a
more diverse pool of species with more equitable abundances (Fig. 6.11).

To explore the influence of speciation rate—via its influence on the regional
species pool—on local diversity, we can simulate local community dynamics
with immigration. With probability m, a local recruit is chosen from the regional pool, which is defined by one of the relative abundance distributions

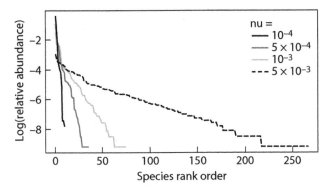

Figure 6.11. Relative abundance distributions in a neutral model of a single community under different speciation rates (nu). In all cases, J = 10,000. Results are shown at a single point in time after 10,000 years. See Online Box 7 for the R code.

in Figure 6.11. A species' relative abundance (i.e., frequency) in the regional pool defines its probability of being chosen to provide the new recruit when an immigration event occurs. This is basically an individual-based version of the island biogeography model (MacArthur and Wilson 1967). The R code for this simulation is shown in Online Box 8.

With these simulations, we see the well-known results that species richness increases with both area (represented by local J) and with the immigration rate, m (Fig. 6.12). We can also see that drawing immigrants from a regional

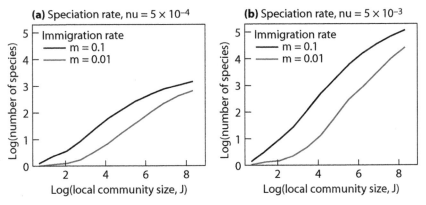

Figure 6.12. Relationships between species richness and community size (i.e., "area") among independent local (and neutral) communities with immigration at different rates (m) and from species pools with different speciation rates (nu). See Online Box 8 for the R code.

pool with a higher speciation rate leads to greater local diversity (Fig. 6.12). This is the crux of the species-pool hypothesis for explaining trends in species diversity along environmental gradients (Taylor et al. 1990): if we imagine that panels (a) and (b) in Figure 6.12 represent two different habitat types (e.g., unproductive and productive), local diversity is greater in (b) owing to a higher regional speciation rate rather than any difference in the nature of local selection in regulating diversity.

6.6. SUMMARY

The different kinds of dynamics described in this chapter might be underlain by innumerable specific mechanisms involving nutrients, disturbance, predators, pathogens, environmental fluctuations, physiological and life-history trade-offs, biogeographic context, and so on. However, these fairly simple scenarios illustrate the key features of a very large number of models describing the dynamics among species interacting on the same trophic level in a local community or in a set of local communities linked by dispersal and ultimately influenced by speciation. Importantly, they can all be generated by altering a few lines of computer code, and they, in turn, generate many predictions for empirical testing (see Chaps. 8–10).

Many researchers already comfortable working with mathematical models will no doubt find these simulations unnecessarily crude, given the availability of analytical models with which the same phenomena can be illustrated, and given certain simplifying assumptions (e.g., constant J) that are not especially amenable to incorporation of certain phenomena of interest (e.g., predator-prey interactions). My target audience in this chapter is the other 95% of ecologists. By converting some simple rules for how things change over time into a language a computer can read, we can generate many predictions for empirical testing while at the same time providing an accessible entry point to understanding the key features of a huge class of ecological models for interacting species. And the underlying high-level processes are strikingly few: drift (as influenced by community size), dispersal, speciation, a few forms of selection, and their variation across space and time.

APPENDIX 6.1. THE RELATIONSHIP BETWEEN FITNESS AND SPECIES' FREQUENCIES

Intuitively, the term *negative frequency-dependent selection* explains itself: the fitness advantage of a given species decreases as a function of that species' frequency. But a monotonic negative function can take many shapes. The qualitative lessons don't depend on the details (within reason), but computer code requires specific instructions, so I am unable to avoid a bit of math here. Because the important parameter is the ratio of species' fitnesses (rather than the difference), a fitness advantage of 1.2 for species 1 (vs. 1.0 for species 2) does not represent a degree of selective advantage equivalent to the disadvantage represented by a fitness of 0.8. In both cases, the fitness difference is 0.2, but the ratios are not the same. In the first case, species 1 has an advantage of $1.2/1.0 = 1.2$, while in the second case species 2 has an advantage of $1.0/0.8 = 1.25$. A fitness ratio of $1.2^{-1} = 0.833$ provides the same advantage to species 2 that species 1 experiences when the fitness ratio is 1.2. To ensure symmetry of this nature in the relationship between fitness and frequency, we can define it using log ratios as follows:

```
log(fit.ratio) <- freq.dep*(freq.1 − 0.5)
                + log(fit.ratio.avg)
```

The "-0.5" serves to define `fit.ratio.avg` as the value of `fit.ratio` when the two species frequencies are equal ($= 0.5$), as well as the average back-transformed logarithm of the fitness ratio averaged across all frequencies. Calculation of the raw `fit.ratio`, which is needed to calculate `Pr.1`, is as follows (see Online Box 2):

```
fit.ratio <- exp(freq.dep*(freq.1-0.5)
              + log(fit.ratio.avg))
```

The relationships in the left panels of Figure 6.3 are slightly curvilinear because `freq.dep` defines a linear relationship with `log(fit.ratio)` rather than with `fit.ratio`. The left panel of Figure 6.5 shows `log(fit.ratio)` on the y-axis and is thus perfectly linear.

PART III
EMPIRICAL EVIDENCE

CHAPTER 7

The Nature of Empirical Evidence

So far, this book has focused largely on concepts, theory, and models. We turn now to empirical studies. Chapters 8–10 will evaluate empirical tests of a series of hypotheses and predictions made by the theory of ecological communities, but first, it is important to cover some philosophical and methodological issues concerning the different ways ecologists go about studying communities empirically. This chapter addresses the following questions:

- How many studies in the literature pertain to the hypotheses and predictions of the theory of ecological communities, and how are these studies distributed according to different taxa and empirical methods?
- What are the main methodological approaches to studying communities empirically, and what are their strengths and weaknesses?
- What motivates empirical studies in community ecology?
- What are the basic units of observation and levels of analyses employed by community ecologists?

Readers already familiar with the various motivations, approaches, and levels of analysis in empirical studies in community ecology may wish to skip to the next chapter. However, I think that reflecting on the tremendous level of heterogeneity among ecological studies (with respect to the factors listed), the scope of the literature, as well as fundamental challenges to our ability to detect cause-effect relationships can be of great assistance in critically evaluating inferences drawn from empirical studies.

7.1. THE STATE OF THE EMPIRICAL LITERATURE

In preparing to write the empirical chapters of this book, I decided to first step out of my "bubble" of knowledge, which is based on a biased subset of what's written on community ecology broadly. As is the case for any ecologist, I know far more about the systems I study (plants in temperate forests and grasslands), and the conceptual or theoretical ideas most pertinent to these empirical studies, than about other systems and topics. My initial foray out of the bubble involved scanning all the papers in one randomly selected issue in each of seven ecological journals in each of the years 2011–2014 (see Box 7.1), from which I learned three key things. First, the topic of this book (horizontal community ecology) is relevant to a large proportion of studies done by ecologists—roughly one-third. There are biases in the literature with respect to taxa and methods, largely in directions one would expect (e.g., lots of observational studies of plants; see Table B.7.1), but I have tried to prevent my own biases from resulting in an overly skewed view of the literature.

The second lesson was that my feeling of being overwhelmed by the sheer quantity of published papers was justified—and then some. Some rough calculations indicate that no fewer than 10,000 papers (and probably at least twice that many) relevant to the topic of this book have been published over the past decade (Box 7.1). I try to avoid exclamation marks in professional writing, but . . . TEN THOUSAND PAPERS! You would need to read no fewer than four new papers every working day of your life just to be able to claim a comprehensive knowledge of this one subtopic of ecology. The many hundreds of papers I did read during the writing of this book were enough to make Google Scholar wonder if I was perhaps a robot (and I didn't always pass the test; see Fig. 7.1), although I certainly came up well short of 10,000. Thus, I made no attempt to comprehensively cover the relevant literature but instead drew on selected examples of tests of particular hypotheses and predictions.

The final lesson—hinted at earlier—was that empirical studies in community ecology are highly heterogeneous, not only with respect to taxa and habitats but also with respect to their underlying motivation, their approach (e.g., observational vs. experimental), and the level at which data are collected and analyzed (e.g., individuals, plots, regions). To make some sense of this heterogeneity, the following sections walk through these different axes that distinguish one empirical study from the next, as well as discuss some widespread challenges faced by empirical ecologists.

7.2. SCIENTIFIC MOTIVATIONS

7.2.1. The Goals of Science: Prediction and Explanation

In addition to describing phenomena of interest, scientists aim to achieve two broad goals: prediction and explanation. Some ecologists have argued that pre-

BOX 7.1.
WHAT'S BEING PUBLISHED IN
COMMUNITY ECOLOGY?

I randomly selected one issue (not including special features) from each of seven journals in each of the years 2011–2014 and read the abstracts of all 500+ papers. I chose respected journals covering the full sweep of ecology: *Ecography, Ecology, Ecology Letters, Global Ecology & Biogeography, Journal of Ecology, Journal of Animal Ecology,* and *Oikos.* Excluding items like brief commentaries on other papers, announcements, errata, and the like, the sample I examined included 502 papers from 28 issues. I classified 173 of these papers (34%) as involving horizontal community ecology. This meant that horizontal community ecology was the primary focus of the paper or one of several topics, or that relevant theory was drawn on to assess the consequences of a community-level property (e.g., the effect of biodiversity on ecosystem function). I scrutinized these 173 papers, recording factors like focal taxon and basic approach.

Eighteen papers were purely theoretical, leaving 155 empirical papers. Almost half of these were about plants, and while essentially all taxa were represented, "higher" plants and animals were much better represented than microbes or algae (Table B.7.1). Roughly one-third involved field or lab experiments, but experiments were more common for some taxa (e.g., plants, invertebrates) than for others (e.g., vertebrates). I found no big surprises in these proportions, but I was taken aback by the astounding number of papers being published.

TABLE B.7.1. Breakdown of 155 Empirical Papers on Horizontal Community Ecology Based on the Focal Taxon and the Basic Approach

	Taxon						
Approach	*Microbes*	*Algae*	*Plants*	*Invertebrates*	*Vertebrates*	*Multiple Taxa*	*Total*
Field observations	5		47	17	20	9	98
Field experiments		4	22	11	1	6	44
Lab experiments	4	1	2	5	1		13
Total	9	5	71	33	22	15	155

Note: If a paper included both experimental and observational components, it was put into one of the experimental categories. A paper including multiple taxa was not also included under the columns for the component taxa (i.e., each paper was assigned a unique category and therefore counted only once).

(Box 7.1 continued)

Some back-of-the-envelope calculations provide a rough answer to the following question: how many papers relevant to the topic of this book have been published over the past decade? The seven journals I examined published a total of 276 issues from 2011 to 2014, so my sample of 28 issues is approximately 10% of the total. We could thus expect to find about 1700 relevant papers in just these seven journals over a period of four years. Accounting for the fact that the number of papers published per journal (on average) tends to increase over time, we would probably find no fewer than 3000 relevant papers in these seven journals over the past 10 years. And with many other journals publishing papers on community ecology, either on particular systems or taxa (e.g., *Journal of Vegetation Science, Marine Ecology Progress Series*), across a wide variety of systems (e.g., *PLoS One, Oecologia*), or in a conservation context (e.g., *Ecological Applications, Journal of Applied Ecology*), I estimate that a bare minimum of 10,000 papers relevant to the topic of this book have been published over the past decade. The true number is probably at least twice as many.

diction is the ultimate currency of science and that one need not have a mechanistic explanation for exactly *why* something will happen to predict that it *will* happen (Peters 1991). For example, the species richness of many taxa at large scales (e.g., >1000 km^2) is highly predictable from the quantity of usable energy, which is a function largely of temperature and precipitation (Currie 1991, Hawkins et al. 2003), and this relationship could be quite useful for making predictions in a changing climate, regardless of why the link exists (Vázquez-Rivera and Currie 2015). Others argue that explanation (i.e., the prerequisite to understanding) is necessary for prediction, such that explanation is the more important goal (Pickett et al. 2007). For example, whatever process created a link between richness and energy in the past might not apply in the future, and we can assess this relationship only by understanding the process (Wiens and Donoghue 2004, Kozak and Wiens 2012).

In Chapters 8–10 I draw on studies of all kinds that can be used to test predictions of process-based hypotheses. These include studies whose primary aim might be statistical prediction given that the direction and strength of such predictive relationships can—at least to some degree—help distinguish between competing process-based hypotheses. In fact, the distinction between prediction and explanation is not so clear-cut. For example, a very strong relationship between species richness and energy that is repeatable across space, time, and taxa points to a strong likelihood that energy falls somewhere on the "true" causal pathway explaining variation in species richness (Currie 1991,

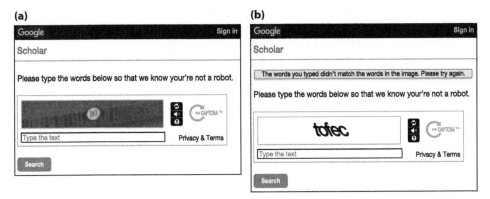

Figure 7.1. (a) What happens in Google Scholar when you access too many papers in a short period of time, and (b) some evidence that the author was in a robotic state while doing so.

Hawkins et al. 2003, Vázquez-Rivera and Currie 2015). In some cases, energy may simply be correlated with the true cause(s) of variation in species richness, but establishing the generality of the pattern still represents a step toward explanation in that we now know that the true cause of richness variation must be something correlated very strongly with energy (Wiens and Donoghue 2004). Correlation does not imply causation of X on Y, but it does imply some causal pathway involving both X and Y (Shipley 2002).

7.2.2. Four Pathways to an Empirical Study

In addition to empirical studies in ecology motivated by prediction versus explanation, there are others motivated by specific questions that fall under at least the following four non–mutually exclusive categories:

1. *What has happened in nature?* This is the most basic ecological question. Many ecological studies quantify intriguing patterns, noted previously only via casual observation (or not at all). Examples include the shape of species abundance distributions (few abundant and many rare species), variation in species diversity along gradients of latitude, disturbance, or island area, and abrupt spatial transitions between areas dominated by different species.

2. *Why have certain things happened in nature?* For the latitudinal gradient in species diversity, we might start by asking which particular environmental or historical variables correlate most strongly with diversity (thereby representing likely causes). In the case of an

abrupt transition between two dominant species along an environmental gradient, we might combine transplant and removal experiments to test whether competition or environmental tolerance play important roles (Connell 1961). We can already see how these different types of questions bleed into one another—simply identifying a strong correlate of some community property (*What has happened in nature?*) represents a step in the direction of figuring out *why* that property varies from place to place.

3. *What are the ecological consequences of factor X?* Our field observations frequently point to particular factors, such as climate, nutrient input, habitat fragmentation, or predation, as having potentially important effects on any number of ecological outcomes. This type of question starts with a causal agent of interest (an independent variable, in statistical parlance), and subsequently asks how exactly some community property (the dependent variable) might be influenced by this factor. In contrast with the previous type of question, which begins with some outcome that has already manifested in nature, here we ask what if? questions, creating or exploring hypothetical scenarios. Such questions are often addressed by experimentally manipulating the factor of interest (e.g., adding predators to a lake, or fragmenting a habitat).

4. *Are the assumptions and predictions of theory Q met in nature?* Many studies in the three previous categories could fall under this category, but other theory-motivated studies do not fall so readily under one of the previous categories. Some empirical tests of coexistence theory are good examples, in which researchers might test for factors like particular kinds of trait or fitness trade-offs among species, or negatively frequency-dependent fitness of individual species (HilleRisLambers et al. 2012). In ecology, data collectors and theory developers have always had a somewhat uneasy relationship, with theory incorporated into empirical studies in various ways (Shrader-Frechette and McCoy 1993). This topic is treated in the next section.

7.2.3. The Role of Theory in Motivating (or Not) Empirical Studies

An idealized view of theory is that it involves one or more hypotheses put forward to explain some nonobvious feature of nature, making specific assumptions and predictions that provide the basis for testing. We can set some guidelines for what should be labeled a theory, model, or hypothesis in ecology (Scheiner and Willig 2011), but in practice, the words *theory* and *hypothesis* have been used to describe more or less any motivation at all for making a prediction about what we might see in nature (Pickett et al. 2007, Marquet et al. 2014).

To my eye, many ecological theories can be summed up as "I predict that you will find in your system what I have found in mine" or "Do you see what I see?" rather than presenting lines of logic that begin by establishing *why* a certain pattern might arise. An example is the intermediate disturbance hypothesis, which has a somewhat controversial history (Fox 2013, Huston 2014). The initial hypothesis worked backward from an observed pattern and involved articulation of some reasons why few species would be found in highly disturbed conditions (few species adapted to such extreme conditions) and in undisturbed conditions (competitive exclusion by one or a few dominant species; Grime 1973). However, "testing" has amounted largely to asking whether the unimodal relationship of diversity versus disturbance that Grime observed in herbaceous vegetation in Britain is also found elsewhere (Fox 2013). While the answer to this question represents an important *empirical* advance (establishing the generality of the pattern once many systems have been studied), I don't think the results feed back in any major way to decide between broader theoretical visions of how the world works. The pattern itself, whether consistent with the hypothesis or not, cannot arbitrate between different possible causes.

Other theories start from first principles, declare one or more assumptions about how the world works, and then use verbal or mathematical logic to develop specific predictions. Ecological neutral theory is an example. Starting from the assumption that individual organisms are demographically equivalent regardless of species, that there is some upper limit on the number of individuals in a community, and that dispersal is spatially limited, Hubbell (2001) developed mathematical models making a series of predictions concerning the accumulation of species with increasing area, the shape of species abundance distributions, and the decay of community similarity with increasing geographic distance. Many studies have explicitly tested the theory's assumptions and predictions (Rosindell et al. 2012, Vellend et al. 2014), and I believe that—unlike for the case of "Do you see what I see?" theories—these results do feed back to inform broader theoretical underpinnings of ecological patterns and processes. For example, strong relationships between species composition and environment are not predicted by neutral theory and thus must involve selection (see Chap. 8), while the consistency of some community patterns with neutral predictions (e.g., the shape of the relative abundance distribution) forces us to rethink whether selection is necessarily involved in their generation (see Chap. 9).

The next three chapters fall more closely in line with the latter use of theory. In Chapters 8–10, a series of hypotheses is stated concerning the importance of a particular process, each of which makes multiple predictions that can be confronted with empirical data. In Chapter 11, I revisit the "pattern-first" approach and explain why and how the theory of ecological communities is equally relevant to studies taking both the process-first and pattern-first approaches.

7.3. FUNDAMENTAL EMPIRICAL APPROACHES: OBSERVATION AND EXPERIMENT

Science begins with observations—of things to explain and of things we'd like to predict. Entirely observational studies are often labeled as "merely descriptive," able to characterize patterns but not to test the action of processes. This point of view lies somewhere between oversimplification and falsehood (Shipley 2002, Sagarin and Pauchard 2012). As astronomers, geologists, and epidemiologists can attest, great progress in understanding process can be achieved via careful observation, development of competing theories and models for observed phenomena, and pursuit of multiple lines of evidence. The same is true of ecology (Pickett et al. 2007). For example, observations of fossil pollen deposited across time and space, glacial history, environmental conditions, and tree demography have been combined with molecular genetics and mathematical models to provide a fairly clear understanding of the processes underlying geographic range shifts of trees and forest community assembly in changing environments (Clark et al. 1998, McLachlan et al. 2005, Williams and Jackson 2007).

When they are possible, manipulative experiments provide an unparalleled tool for testing targeted hypotheses or simply for understanding better how a system works (Hairston 1989, Resetarits and Bernardo 1998, Naeem 2001). For example, to test the hypothesis that local species diversity is limited by the rate of immigration, there is no more direct way than experimentally increasing the rate of immigration of new species (Turnbull et al. 2000). Experiments can be conducted in the field (e.g., adding seeds to a natural plot of vegetation), in the lab (e.g., connecting microcosms via dispersal to different degrees), or in settings that combine elements of the lab and field (e.g., setting up containers outdoors to resemble miniature ponds).

Despite the obvious importance of experiments, they are not a panacea for ecological science, as mentioned briefly in Chapter 3. In addition to being infeasible or unethical in many scenarios, experiments are severely restricted in the spatial scale at which they can be implemented, and experimental manipulations are often of uncertain applicability to unmanipulated nature (Bender et al. 1984, Yodzis 1988, Dunham and Beaupre 1998, Petraitis 1998, Werner 1998, Maurer 1999, Naeem 2001). For example, one can warm a plot of ground or a small water body to test the ecological effects of temperature, but the experimentally imposed temperature change almost always occurs faster than it would naturally and is exceedingly difficult to implement without simultaneously altering other factors. One would not want to rely only on experimental manipulations of temperature to make predictions of how ecological communi-

TABLE 7.1. Some Advantages and Disadvantages of Different Empirical Ecological Studies in Terms of Their Realism, Their Ability to Elucidate Processes Underlying Particular Patterns or Outcomes, and Whether "People Care"

Approach	Realism	Process	People Care
Observational: bird community properties in naturally fragmented and unfragmented landscapes	High	Low	High
Observational: successional trajectories in forests with different land-use histories	High	Low	High
Experimental: effects of predator removal/addition in a field "container community" (e.g., the water in tree cavities)	High	High	Low
Experimental: effects of nutrient addition on zooplankton in cattle tanks	Medium	High	Medium
Experimental: plant community response to manipulation of temperature in the field	Medium	Medium	High
Experimental: creation of experimental communities in the field to assess the influence of diversity on productivity, or the importance of particular mechanisms of coexistence	Medium	Medium	Medium
Experimental: effects of dispersal in lab microcosms of microbes	Low	High	Low

Note: *Realism* is the degree to which the results apply to nonexperimental situations. Whether "people care" indicates whether nonecologists in the general public or potential end-users in management or policy circles are likely to take an interest in the results.

ties will respond to global warming (Wolkovich et al. 2012). Finally, although not a strictly scientific criterion, the feasibility of ecological experimentation is typically lowest for response variables that the general public cares about the most, such as the diversity of large mammals or birds.

Many observational studies have been dubbed "natural experiments" (Diamond 1986), describing situations in which some kind of human-caused or natural environmental change has occurred whose consequences can be analyzed via observational study. For example, a disturbance may have occurred in some areas but not others (e.g., fire, logging, or exotic species invasion), thus mimicking what one might like to have done with a "true" experimental manipulation. Just about any observational study examining sites under different conditions could be considered a natural experiment, although the degree to which a factor of interest (e.g., a particular disturbance) has occurred independently of potentially confounding factors (which can be controlled in a manipulative experiment) varies tremendously from one study to the next. Nevertheless, taking advantage of natural experiments in nature is certainly worthwhile.

In sum, different approaches clearly have numerous advantages and disadvantages relative to one another. Table 7.1 presents a subjective evaluation of different types of ecological studies— meant to be representative of the broad sweep of community ecology—in terms of their realism, ability to elucidate mechanism, and whether people generally care about the results. In Chapters 8–10, I take a highly pluralistic approach, drawing on empirical studies of any type that speak to predictions made by different hypotheses.

7.4. LEVELS OF ANALYSIS IN COMMUNITY ECOLOGY

The substantial variation among ecological studies according to the factors just described—overarching goals, sources of motivation, and empirical approaches —is dwarfed by variation among studies in their units of observation, levels of analysis, and the ecological variables of interest.

The *unit of observation* is the object on which measurements of interest are made. In ecology these might be organs, individual organisms, populations, species, quadrats, or entire regions. *Levels of analysis* define the units used in a particular analysis. While it might seem self-evident that analyses in community ecology target the community level, in fact, theory in community ecology makes predictions at several levels (Table 7.2). Some studies have as their level of analysis the individual tree or seedling, with which they test predictions from community ecology theory concerning variation in survival or growth as a function of the density of conspecifics and heterospecifics (Comita et al. 2010). For example, in the world-famous 50-ha tropical forest plot on Barro Colorado Island (BCI), Panama, the basic unit of observation of long-term monitoring is the individual tree (sometimes seedling), with the key measurements or observations being the tree's species identity, size, and geographic position (Hubbell and Foster 1986). Another possible level of analysis is the species, in which case one might aggregate data across individuals of each species to test for trade-offs (between traits or fitness components) that are predicted by theory to promote stable coexistence (Wright et al. 2010).

Aggregating above the level of individual trees, scientists can study plots or "local communities" of any size (e.g., 50 × 50 m) and at this level of analysis test for composition-environment or diversity-environment relationships (John et al. 2007). Finally, researchers can draw on data from similar plots in different regions, making an entire 50-ha plot the level of analysis, and allowing them to ask, for example, how within-plot beta diversity (compositional variation among subplots) or any other property varies according to environmental conditions (De Cáceres et al. 2012). As described in Chapter 2, the relationship between a site property (e.g., environment) and a first-order community pattern (e.g., species diversity) can itself be thought of as a community pattern, in this case a "second-order" pattern. Thus, we could also ask how the strength of the relationship between topography and species composition varies among different forest plots (De Cáceres et al. 2012) or among different subsets of the

TABLE 7.2. Levels of Analysis Used in Empirical Studies in Community
Ecology, and Properties of Interest at Each Level

Level of Analysis	Properties to Explain (i.e., Y-, or Dependent Variables)	Potential Explanatory Properties (i.e., X-, or Independent Variables)
1. Site / local community	• Diversity • Composition (possibly involving traits in either case)	• Environment (mean, variance) • Size/area • Surroundings (e.g., connectivity) • Community subset (e.g., an exotic species) • Time/age • Site history
2. Pair of sites	• Compositional dissimilarity (pairwise beta diversity)[a]	• Geographic distance • Difference in any site-level property

3. Set of sites (>2, e.g., in a region[b] or experimental treatment)	• Overall beta diversity
	• Any relationship between properties listed in rows 1 and 2
		• Regional- or metacommunity-scale estimates of properties listed above, such as
		  • Regional species diversity
		  • Distribution of species or traits in the regional pool
		  • Environment (e.g., climate)

4. Individual or population	• Fitness/performance • Traits	• Density/frequency of conspecifics and other species • Environment
5. Species	• Fitness (overall or a particular component) • Presence/abundance • Trait	• Other fitness components (not used as a Y variable) • Presence/abundance of other species • Other traits
6. Pair of species	• Niche difference • Fitness difference • Interaction strength/direction	• Phylogenetic or trait difference • Environment
7. Set of species (>2)	• Any Y variable or X-Y relationship listed above	• Average trait • Shape of tradeoff between multiple traits

Note: The term *environment* is used here as a catch-all for any site property that can be measured
independently of the focal community itself (see Fig. 2.1e); common examples include levels of
resources used by the focal community (e.g., nutrients, prey), abiotic conditions such as temperature or pH, disturbance regime, and consumer pressure (e.g., herbivory, predation).

[a]For pairwise analyses, any univariate site/species property can also be used to calculate a pairwise difference, but this is redundant with analyses at the site/species level, so these properties
are not listed here. Pairwise differences in multivariate space (e.g., species composition, geographic coordinates) cannot be reduced to single axes without loss of information.

[b]By *region* I mean whatever area encompasses a set of study plots, which can be any physical size.

community analyzed separately (e.g., slow-growing vs. fast-growing species). In the latter case, the level of analysis is a set of species.

In many studies, the unit of observation is not an individual organism but rather a study plot of a given size. For example, ecologists often record percent cover as an estimate of plot-level abundance for species of plants or sessile invertebrates. In these cases, the plot-level data cannot be disaggregated into smaller units of observation, although one does have the option of considering species as the units of observation and asking questions about correlated distribution patterns among species (Gotelli and Graves 1996, Legendre and Legendre 2012). Plot-level data can, however, be aggregated at different levels of analysis. As with the BCI data, one can examine variation in community properties among plots, among pairs of plots, or among entire sets of plots (e.g., those subject to different experimental treatments). Finally, it is important to note that the concept of levels of analysis is only very loosely related to the issue of spatial scale. Study plots can be 1-mm^3 samples of soil or hundreds of square kilometers of land or sea, with the only scale constraint being that multiple plots take up more space than single plots (within a given study).

Depending on the level of analysis, researchers focus on a variety of different community properties they would like to explain or predict (i.e., the dependent variables in an analysis) and an even larger number of possible explanatory or "independent" variables (Table 7.2). While it may be difficult to keep track of all such questions explored by community ecologists, considering the unit of observation and level of analysis used to pose a particular question can help them figure out how one study relates to the next—often a nontrivial challenge in ecology. Empirical studies discussed in the next three chapters involve many combinations of units of observation and levels of analysis.

7.5. CONFOUNDING VARIABLES AND INFERRING X-TO-Y CAUSATION: THE THREE-BOX PROBLEM

Regardless of the motivation, approach, or level of analysis, community ecologists are constantly faced with the problem of potentially confounding variables. Most if not all predictions flowing out of community ecological theory involve variation in some biotic property Y (e.g., individual fitness, community composition) being caused by some other abiotic or biotic variable X (e.g., conspecific frequency, environmental conditions). A widespread challenge to directly testing such predictions is the presence of many other variables that covary with X and Y and that may thereby create or magnify X-Y correlations, despite weak or nonexistent causal effects. Many types of empirical study must confront this challenge (Fig. 7.2).

There are two main solutions. First, for observational studies we can attempt to measure the most likely confounding variables and subsequently "control"

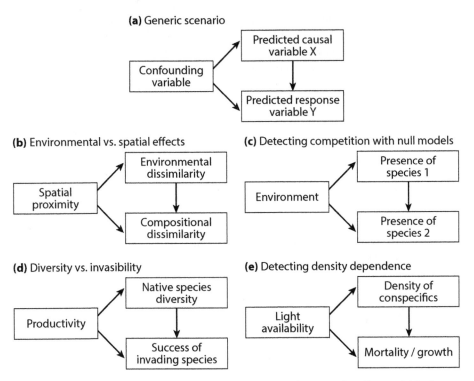

Figure 7.2. The "three-box problem" in community ecology: confounding variables in attempts to detect cause-effect relationships. Panel (a) illustrates the general challenge, with specific manifestations illustrated in (b)–(e). For simplicity, I have not indicated the possibility of an effect of *Y* on *X*, or of the confounding variable being influenced by *X* or *Y*, both of which are possible.

for them statistically when assessing particular *X-Y* relationships. This is most commonly done in one of two ways: (i) Using linear models of one sort or another in which *Y* is predicted as a function of X_1, X_2, and so on, with the effect of each predictor X_i evaluated after controlling for the effect of the other *X*'s; (ii) using path analysis or structural equation modeling, which accounts for more complex cause-effect relationships among a set of variables (Shipley 2002). The risk in such observational studies is failing to identify (and therefore measure) important confounding variables.

The other solution is to experimentally manipulate the factor of interest, with appropriate control of confounding variables via replication and randomization. This approach permits the most direct control of potentially confounding variables, but it is not without limitations. As already noted, experiments may not be possible for particular systems or spatial scales, and all manipulations of *X* are of at least somewhat uncertain relevance to interpreting *X-Y* relationships in unmanipulated nature. In addition, manipulation of one variable (e.g., species richness, spatial resource heterogeneity) might involve the unwitting simultaneous manipulation of other variables (e.g., species composition, maximum microsite resource levels). The latter problem may often be solved via improved experimental design, although the logistical challenges can be considerable. In the next two chapters on empirical evidence, we will repeatedly face the three-box problem.

7.6. THE LITERATURE IS A BIG, HETEROGENEOUS LANDSCAPE

To summarize, the ecological literature is vast—ecologists are motivated by a variety of goals, they employ many different approaches with different units of observation and levels of analysis, and they frequently need to deal with potentially confounding variables. This is admittedly a bit of a grab bag of issues, but the overarching theme is that there are many ways to do empirical community ecology, and each decision made during the design of an empirical study brings with it various advantages and disadvantages. Thus, the stage is set for the remainder of the book, in which we will encounter a great diversity of different types of empirical study in community ecology.

CHAPTER 8

Empirical Evidence: Selection

The next three chapters pursue the following question: what is the empirical evidence for selection, drift, dispersal, and speciation in determining community structure and dynamics? The current chapter covers different forms of selection, Chapter 9 covers drift and dispersal, and Chapter 10 covers speciation.

In each of these chapters, I first state a hypothesis about the importance of a particular process, which is followed by one or more predictions made by the hypothesis, all of which derive from verbal or quantitative models presented in Chapters 5 and 6. For each prediction, I then briefly describe the empirical methods used to test it, the results of empirical tests, and some limitations of the methods. For the most part I describe only a handful of case studies for each prediction, although when possible I assess the overall strength of empirical support across different studies or systems. To do so, I draw on meta-analyses or review papers when available, and otherwise I just describe my qualitative impressions of the literature. For each hypothesis, I also briefly describe frequently asked questions (FAQ) concerning the low-level processes or factors that underlie the high-level process of interest. I then end each chapter with a tabular summary of the evidence for each hypothesis or prediction, and associated challenges and caveats.

As illustrated in Chapter 5, empirical tests of high-level processes (especially selection) have been described using a large and ever-growing set of terms (see Table 5.1). The following three chapters summarize the vast empirical literature in community ecology using a simplified, hierarchical set of terms based on the theory of ecological communities. Throughout these chapters I mention some of the more common synonymous terms used to describe certain hypotheses or predictions, but I make no attempt to be comprehensive. Table 5.1 can be consulted for many of the most common such terms and their mapping onto the approach taken here.

8.1. HYPOTHESIS 1: CONSTANT SELECTION WITHIN SITES AND SPATIALLY VARIABLE SELECTION AMONG SITES ARE IMPORTANT DETERMINANTS OF COMMUNITY STRUCTURE AND DYNAMICS

I consider these two forms of selection simultaneously because spatially variable selection is, to a large extent, the manifestation of different strengths and directions of constant selection (i.e., selection favoring particular species, regardless of their frequencies) within sites. A key assumption here is that some aspect of the abiotic or biotic environment is the cause of selection, favoring different species in different sites.

Prediction 1a: Species composition is correlated with environmental conditions (abiotic or biotic) across space.

Methods 1a: In a set of study plots in the field, quantify species composition and measure environmental variables thought to underlie selection.

Studies of this nature are as old as ecology itself, although the jargon used to describe them has changed over time, and there has been a steady increase in quantitative sophistication. "Gradient analyses" of the 1960s–1980s (Whittaker 1975) are now often described as tests of "species sorting" (Leibold et al. 2004). Most contemporary studies use multivariate statistics of one sort or another for data analysis. In essence, the vector of species abundances at a given site is considered as a multivariate response, and regression-type analyses are used to test for significant predictive ability of various environmental variables (Legendre and Legendre 2012). Many such analyses have equivalent versions that begin with calculation of pairwise indices of community differentiation (beta diversity), then subsequently assess the predictive capacity of environmental variation among sites via "distance-based" analysis (Anderson et al. 2011). Over the past 20 years or so, a major effort has aimed to ensure that environment effects on species composition are not confounded with effects of site-to-site spatial proximity (Fig. 7.2b). Specifically, if there is spatial autocorrelation of properties of the abiotic environment (Bell et al. 1993) and of species composition (e.g., due to spatially limited dispersal), overlaying the two independent patterns can lead to statistical detection of composition-environment correlations despite the lack of any causal relationship (Legendre and Fortin 1989).

Results 1a: That species composition is correlated with environmental conditions is glaringly obvious, even before quantitative data are collected. Field guides to any group of organisms are replete with descriptions of species' habitat affinities, which combine to create composition-environment relationships.

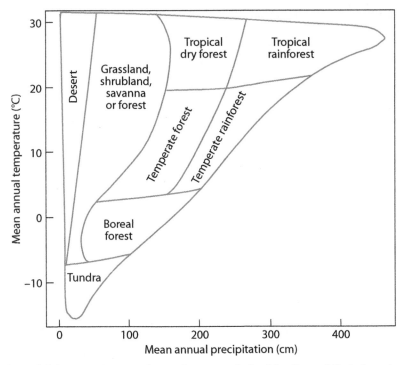

Figure 8.1. Large-scale vegetation-environment relationships. In spatially independent regions, one finds similar (albeit not identical) relationships between temperature, rainfall, and plant species composition, expressed here as biome type. At low (but not the lowest) precipitation, nonclimatic factors such as soil, herbivory, and fire play a major role in determining whether vegetation is grassland, shrubland, savanna, or forest. Modified from Whittaker (1975).

Climatic variables, for example, are strongly predictive of broad-scale patterns of community composition in terrestrial vegetation (Fig. 8.1) and associated communities of animals, such as birds (Fig. 8.2). Not surprisingly, the vast majority of data sets analyzed quantitatively to control for spatial proximity still show clear evidence of composition-environment relationships (Cottenie 2005, Soininen 2014). Composition-environment relationships apply to scales and taxa ranging from microbes sampled in a few cubic millimeters of substrate (Nemergut et al. 2013) to the vegetation-defined biomes of the earth (Merriam 1894, Whittaker 1975).

Predictions 1b: (i) The range or variance in trait values in a set of locally co-occurring species is less than that in random selections of species from the regional pool. (ii) Mean plot-level trait values are correlated with environmental conditions.

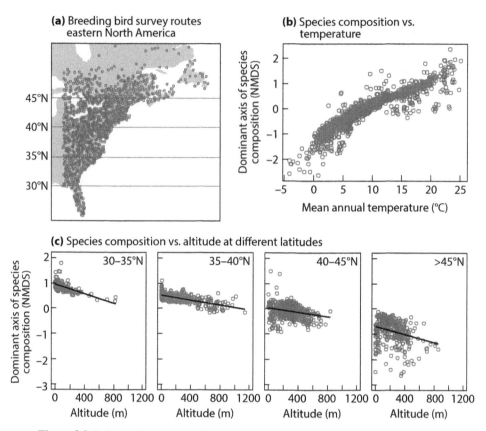

Figure 8.2. Relationships between bird species composition, mean annual temperature, and altitude in eastern North America. For all passerine birds (177 spp.) east of longitude 85°W, I calculated the average abundance across years (up to 2012) for 1548 routes in the North American Breeding Bird Survey (USGS 2013). A nonmetric multidimensional scaling (NMDS) analysis using the Bray-Curtis index was conducted to identify the dominant axis of species composition. The strong relationship between species composition and temperature across the entire region (b) is also reflected in relationships with altitude within subregions (c): cold high-altitude sites have low values along the composition axis. Lines in (c) show best-fit linear least-squares regressions.

A key assumption here is that some of the measured traits mediate composition-environment relationships. For example, possessing high values of a given trait (e.g., body size) might confer high fitness and thus permit high abundance under some environmental conditions (e.g., low temperature) but not others (e.g., high temperature). Results supporting these predictions are often described as indicative of "habitat filtering" or "environmental filtering" (Cornwell et al. 2006, Kraft et al. 2008, Cornwell and Ackerly 2009), although for

Prediction 1b-(i) we need not know in advance which environmental variables are the strongest predictors of community composition. Mean trait values are essentially one way to quantify the "trait composition" of a community, and thus Prediction 1b-(ii) could also be considered an offshoot of Prediction 1a.

Methods 1b: In a set of study plots in the field, measure species-level traits, observe community composition, and calculate plot-level trait statistics (mean and range/variance). For Prediction 1b-(i), compare the observed range/variance with expectations based on a null model (see next paragraph). As a special case, species have been compared based on their phylogenetic similarity rather than their trait similarity, under the assumption that one or more traits important to determining species' environmental affinities are evolutionarily conserved (Webb et al. 2002), although this assumption is questionable (see results 2c). For Prediction 1b-(ii), assess correlations between mean traits and environment.

Since each local community invariably contains fewer species than the regional pool, we cannot simply compare the raw ranges of local and regional trait values. Let's say that local communities contain on average 10 species out of a regional pool of 100. Any sample of 10 species from the regional pool necessarily covers a range of trait values that is less than or equal to the range in the regional pool. Thus we need a null model (Gotelli and Graves 1996). For a given site with S species, the typical approach in this case is to make repeated random draws of S species from the regional pool, with the probability of choosing each species weighted according to the number of sites in which it actually occurs. For each random draw, we calculate the range of trait values and thus generate a "null" distribution of trait ranges, with which we can assess whether the observed trait range is significantly smaller (i.e., smaller than 95% of the values in the null distribution).

Results 1b: Both these predictions enjoy widespread support in nature (Weiher and Keddy 1995, Kraft et al. 2008, Cornwell and Ackerly 2009, Vamosi et al. 2009, Weiher et al. 2011, HilleRisLambers et al. 2012). Plants are especially well represented in such studies (Kraft, Adler, et al. 2015), with the most commonly measured plant trait likely being specific leaf area (SLA, the ratio of leaf area to dry mass), which is correlated with leaf life span and photosynthetic rate (Wright et al. 2004). In a shrubland in coastal California, Cornwell and Ackerly (2009) found increasing mean SLA with increasing soil water content, as well as significantly narrower ranges of SLA values than predicted by a null model (Fig. 8.3). Other studies with similar results cover a wide range of taxa and include analyses of body size in mammals (Rodríguez et al. 2008), tongue length in bumble bees (Harmon-Threatt and Ackerly 2013), and colony morphology of corals (Sommer et al. 2013). It is important to note that while the sophistication of quantitative analysis has advanced tremendously in recent years (e.g., van der Plas et al. 2015), qualitative observations consistent with these predictions go back many decades (Tansley 1939, Margalef 1978, Grime 1979, Weiher and Keddy 1995).

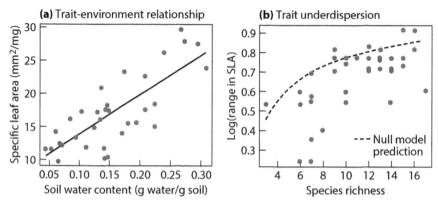

Figure 8.3. Support for Prediction 1b in 44 woody plant communities (20 × 20 m plots) in coastal California. The abundance-weighted average specific leaf area (SLA) across species is strongly correlated with soil water content (a), and the range of local SLA values tends to be smaller than expected based on random selections of species from the regional species pool (b: most data points lie below the null-model prediction). The line in (a) shows the best-fit linear least-squares regression. Data from Cornwell and Ackerly (2009).

Prediction 1c: Altering environment conditions will cause nonrandom changes in species composition.

If composition-environment relationships are created via spatially variable selection, then if we alter local selection experimentally (i.e., we change some aspect of the environment), we should expect predictable shifts in community composition. Since environmental conditions and community composition are always changing to some degree, testing this prediction requires simultaneously observing community changes (or the lack thereof) in sites where an environmental change of interest has not occurred—that is, control sites.

Methods 1c: Experimentally alter environmental conditions (e.g., temperature, resources, presence of a predator or pathogen); observe changes in species composition in altered and unaltered sites.

Results 1c: Support for this prediction is very strong: environmental manipulations invariably cause some degree of change in community composition. Classic two-species lab experiments with plants on different soil types (Tansley 1917), beetles under different conditions of temperature or humidity (Park 1954), and phytoplankton under different nutrient supply rates (Tilman 1977) have shown switches in which species come to dominate depending on the manipulated conditions. In the field, researchers have demonstrated community-level responses to experimental manipulation of many potentially important factors: climate variables, soil/water chemistry, disturbance, predators or her-

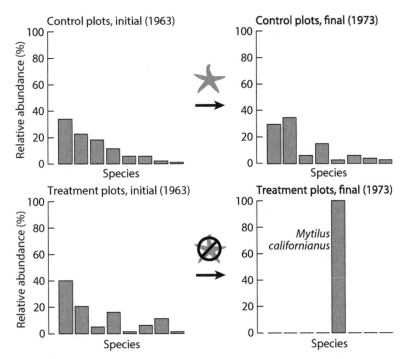

Figure 8.4. Effects of a predator (*Pisaster ochraceus*, the purple sea star) on a marine intertidal community of barnacles (species 1, 4, and 8), mussels (species 5), algae (species 2, 3, and 7), and sponges (species 6). Species are ordered on the *x*-axis according to their relative abundance in the control plots at the start of the experiment. Removal of the sea star caused monodominance by the California mussel (*Mytilus californianus*). Data from Paine (1974).

bivores, mutualists (e.g., mycorrhizal fungi), and so forth (Ricklefs and Miller 1999, Gurevitch et al. 2006, Krebs 2009). For example, in a classic experiment, Paine (1974) demonstrated massive changes in intertidal community composition due to predator removal (Fig. 8.4).

Prediction 1d: Species diversity increases as a function of spatial environmental heterogeneity.

Methods 1d: (i) Observe communities in sites that vary naturally in their degree of environmental heterogeneity. (ii) Create experimental environments with different degrees of spatial environmental heterogeneity; observe community responses.

Spatial environmental heterogeneity is intuitively straightforward but is quantitatively represented in many ways (Kolasa and Rollo 1991). I conceive of it here as follows. Within a given study plot, measurements of environmental conditions can be taken at multiple spatial positions or "microsites" (in two or three spatial dimensions). Quantifying some aspect of the variability or diversity of these measurements represents spatial environmental heterogeneity, which can be based on variance of a continuously distributed variable (e.g., pH), the number of types for a class variable (e.g., soil types), or the evenness of how often each of several types occurs (e.g., the amount of foliage in different categories of canopy height). Scale is obviously of paramount importance here: spatially variable selection at a relatively large scale results from constant selection at a small scale.

Results 1d: Support for this prediction is mixed. Most observational studies have found positive correlations between species diversity and environmental heterogeneity (reviewed by Tews et al. 2004, Lundholm 2009, Stein et al. 2014). A classic example is MacArthur's demonstration of greater species diversity of birds in forests with greater vertical heterogeneity of the tree canopy (MacArthur 1964) (Fig. 8.5a), with similar results found for lizards by Pianka (1967) (Fig. 8.5b). Other examples can be found for animals and plants of various kinds in terrestrial, freshwater, and marine environments, although some studies find no correlation or even a negative correlation between diversity and heterogeneity (reviewed in Tews et al. 2004, Lundholm 2009, Tamme et al. 2010, Stein et al. 2014). Relatively few studies have experimentally manipulated environmental heterogeneity in the field, perhaps owing to logistic challenges of doing so at relevant spatial scales. Most such studies have focused on plants, and the range of results (relationships mostly positive, some nonsignificant or negative) is quite similar to those of observational studies (Tamme et al. 2010).

As with other studies involving environmental variables (e.g., Prediction 1a), there is always the possibility that tests of the diversity-heterogeneity prediction fail to include the environmental variables to which the focal community responds most strongly. Such studies are also quite susceptible to the confounding factors. Imagine a set of 1-ha plots of some terrestrial ecosystem. Within each plot there might be different degrees of spatial heterogeneity in soil nutrient and water content (and thus potential productivity), with consequent effects on communities of plants, animals, or microbes. Across the broader landscape, some conditions (e.g., high productivity) are quite likely to be much more common than others (e.g., low productivity), such that the relatively homogenous plots might be of uniformly high productivity while the heterogeneous plots contain a mixture of high- and low-productivity microsites. In this case, heterogeneity will be confounded with average conditions (Tamme et al. 2010, Seiferling et al. 2014): average productivity is lower in the heterogeneous plots than in the homogeneous plots. Average conditions (in this case productivity) might, in turn, be an important determinant of diversity via their influence on other forms of selection, drift, or historical speciation. In this same scenario, the microsites of a given type (high vs. low productivity) would also be more fragmented in the heterogeneous plots, with consequent effects of "microfrag-

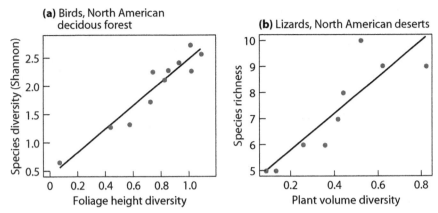

Figure 8.5. Positive relationships between species diversity and environmental hetero-geneity. (a) Birds in forests, with species diversity calculated using the Shannon diver-sity index. (b) Lizards in desert habitats, with species diversity expressed as the number (richness) of species. The *x*-axis values were calculated using the Shannon diversity index, expressing the evenness of abundance of vegetation in three foliage heights (a) or three plant sizes or "volumes" (b). Lines show best-fit linear least-squares regressions. Data from (a) MacArthur and MacArthur (1961) and (b) Pianka (1967).

mentation" via drift (Laanisto et al. 2013). Experimental studies can attempt to control such confounding factors, although rarely in such a way as to directly address the field situations of interest (e.g., hectares of forest).

FAQ FOR LOW-LEVEL PROCESSES UNDERLYING SPATIALLY VARIABLE SELECTION

Which environmental variables underlie selection? In observational studies, ecologists frequently assess the ability of multiple environmental variables to explain site-to-site variation in species composition, although one must always remain cognizant of the possibility that correlated but unmeasured variables are the "real" causal agents. In some cases, observational studies are followed by experimental assessments of either how altering a focal variable of interest influences communities in the field or how different species respond to envi-ronmental variation in the lab (e.g., Tilman et al. 1982, Litchman and Klaus-meier 2008). To test the likelihood that a particular environmental driver is responsible for constant selection, one can also make a priori predictions of which species will increase or decrease in abundance over time. For example, with climate warming, the prediction is that warm-adapted species (e.g., those distributed largely in relatively warm areas) will increase, and cold-adapted species (e.g., those distributed in cold areas) will decrease (Devictor et al. 2012, De Frenne et al. 2013).

Abiotic variables: direct effects on fitness or indirect effects via competition?
Even if both observational and experimental evidence align to indicate that a
given abiotic variable (e.g., temperature) is a dominant influence on community dynamics, the effects of the abiotic variable can occur via at least two
distinct pathways. First, in the absence of interactions with other species, abiotic
conditions can directly influence fitness, preventing the occurrence or reducing the
abundance of a particular species in a subset of conditions (e.g., warm sites). Alternatively, a species might be prevented from occurring in warm sites because a particular antagonist (e.g., a competing or predatory species) lives only in those sites.
For example, classic transplant and removal experiments with barnacles demonstrated an important role for competition, rather than prolonged submergence, in
determining the lower depth limit of species distributions (Connell 1961).

Which traits mediate the response to selection? Similar to studies involving multiple environmental variables, trait-based studies often include multiple
traits. Those traits showing the strongest correlations with environmental variables or the largest deviations from null expectations are inferred to play an important role in mediating the community response to selection. The possibility
always remains that the measured traits are simply correlates of those traits that
"really" mediate selection, although this is quite often explicitly recognized
when quantifying "soft" (i.e., easy to measure) traits for the very reason that
they are correlates of "hard" (i.e., difficult to measure) traits, such as physiological rates (Hodgson et al. 1999, Violle et al. 2007).

8.2. HYPOTHESIS 2: NEGATIVE FREQUENCY-DEPENDENT SELECTION IS AN IMPORTANT DETERMINANT OF COMMUNITY STRUCTURE AND DYNAMICS

This hypothesis is closely linked to species coexistence theory (Chesson 2000b,
Siepielski and McPeek 2010), but it is not equivalent for several reasons. First,
negative frequency-dependent selection is a necessary but not sufficient condition for stable species coexistence (see Chap. 6). Second, even if negative
frequency-dependent selection is not of sufficient strength to counter constant
selection and lead to stable coexistence (Fig. 6.3c), it can contribute to the
persistence of a species in a given place if, for example, there is also dispersal
from other places where the species has higher fitness; it can also have a major
impact on transient community dynamics (i.e., while the community is not
at equilibrium; Hastings 2004, Fukami and Nakajima 2011). Finally, the oft-
cited "invasibility criterion" for coexistence—positive population growth rates
when each species is reduced to extremely low abundance, and other species
abundances have come to a new equilibrium (Chesson 2000b)—is not actually
a prerequisite for the existence of a stable equilibrium point at which multiple species can coexist (see Box 8.1). In short, empirical studies of negative

BOX 8.1.
NEGATIVE FREQUENCY-DEPENDENT
SELECTION, INVASIBILITY, AND COEXISTENCE

One diagnostic signature of stable species coexistence has been termed the "invasibility criterion" (Chesson 2000b). The invasibility criterion is exceedingly difficult to test empirically because it involves reducing each species (one at a time) to a low enough density that it has no influence on other species, then allowing the other species in the community to arrive at new equilibrium abundances, and finally assessing the population growth rate of the focal species that has been "let go" (Siepielski and McPeek 2010). If these growth rates are positive for all species (i.e., invasibility is mutual), coexistence is considered stable. However, a more general definition of a stable equilibrium is any state (e.g., two species' frequencies) for which there is a tendency to return to the same state following perturbations. Thus, it is entirely possible that two species coexist stably given perturbations over a wide range of community states, even if this range excludes very low densities of one or the other species. For example, species 1 (but not species 2) might be subject to an Allee effect (Allee et al. 1949), because difficulty in finding mates at low density causes reduced fitness and competitive exclusion by species 2. The end result is a situation in which stable species coexistence is robust to fairly large perturbations despite failure of the invasibility criterion for one of the two species (Fig. B.8.1).

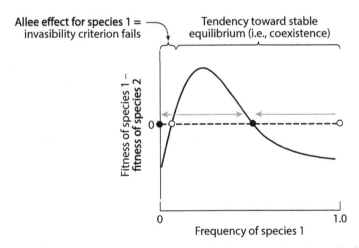

Figure B.8.1. A hypothetical scenario of complex frequency-dependent selection in which species 1 has low fitness at extremely low frequencies (e.g., owing to an Allee effect), but otherwise, fitness in both species is negatively frequency dependent across a wide range of community states. As in Chapter 5, filled circles represent stable equilibria, open circles unstable equilibria, and gray arrows the expected direction of community change.

frequency-dependent selection are of importance beyond testing the strict conditions of coexistence theory.

The following predictions include both general tests of (likely) importance and also a test of whether negative frequency-dependent selection is *important enough* to lead to stable coexistence. Note that the studies in this section pertain largely to selection emerging from local-scale interactions in relatively homogenous environments, rather than from spatial heterogeneity, the latter of which (addressed under hypothesis 1) can also be thought of as negative frequency-dependence emerging at a larger spatial scale (Chesson 2000b).

Prediction 2a: Fitness is more negatively affected by intraspecific density than by the density of other species.

A relative advantage when rare (negative frequency dependence) implies that decreasing a species' abundance results in a "release" from strong intraspecific competition. This prediction is thus often phrased as "intraspecific competition is stronger than interspecific competition." However, it is typically assessed by measuring species' responses to altered densities of conspecifics and heterospecifics, and, as described in the FAQ later, density- or frequency-dependent interactions between species on the same trophic level can be mediated by many low-level processes (Dunham and Beaupre 1998). These processes include competition via direct interference, production of toxins, or exploitation of common resources, "apparent competition" (Holt 1977) via shared predators, pathogens, or mutualists (e.g., plant-soil feedbacks), and facilitation (Ricklefs and Miller 1999, Krebs 2009). Therefore, the high-level Prediction (2a) does not require direct competition per se. A corollary prediction is that individual performance should be greater in a community of multiple species than in a monoculture of just one species.

Methods 2a: (i) Create de novo experimental communities with varying densities and frequencies of different species, (ii) experimentally modify the density or frequency of species in field communities, or (iii) observe communities that vary in composition over space and/or time. In each case, quantify fitness, components of fitness (e.g., growth), or changes in species abundances over time.

Results 2a: Support for this prediction varies greatly from study to study. Hundreds if not thousands of experiments have modified the densities and/or frequencies of potentially competing animals, plants, fungi, or microbes in one way or another (Connell 1983, Schoener 1983a, Goldberg and Barton 1992, Gurevitch et al. 1992). The vast majority of such studies focus on relatively short-term responses and only one or a few components of fitness or population growth (e.g., biomass change over one season or year). One early review concluded that, as predicted, intraspecific effects were more strongly negative than interspecific effects (Connell 1983). However, other reviews and meta-analyses found no such general tendency (Schoener 1983a, Goldberg and Barton 1992, Gurevitch et al. 1992), even if some individual studies do support this prediction (Fig. 8.6).

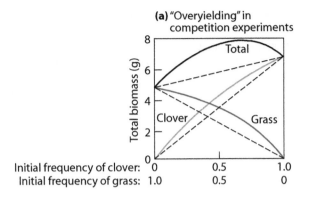

(a) "Overyielding" in
competition experiments

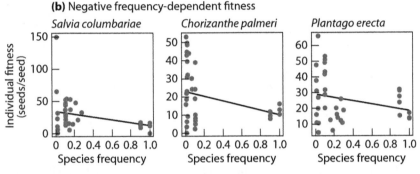

(b) Negative frequency-dependent fitness

Figure 8.6. Evidence for stronger intra- than interspecific negative effects on fitness. (a) Greater biomass production in mixtures of two species than in monocultures of each. Redrawn from Jolliffe (2000); the two species are *Trifolium repens* (clover) and *Lolium perenne* (grass), collected from a 16-year-old pasture and grown at a total density of 24 plants per 13-cm-diameter pot. Solid lines are fits to data; dotted lines are expectations for equal intra- and interspecific effects. (b) Per capita population growth (fitness) as a function of initial frequency for the three most abundant species in experimental communities of California annual plants. Data in (b) are from Levine and HilleRisLambers (2009). Lines in (b) show best-fit linear least-squares regressions.

Many studies over the past 25 years with a variety of plants, herbivores, and predators have experimentally manipulated initial species abundances, not with the explicit goal of testing Prediction 2a but rather to test for effects of species richness on total community-level abundance or productivity (Cardinale et al. 2012). However, the results do address this prediction indirectly. Most often, such studies find that species-specific performance (most often biomass accumulation relative to initial abundance) is, on average, greater in experimental communities with multiple species than in single-species monocultures (Cardinale et al. 2012). Part of this effect can result from diverse communities simply being more likely to contain the "best" species (Aarssen 1997), but complementarity among species (i.e., the general tendency for species to perform better when in a diverse community) appears to play an important role (Cardinale et al. 2007). These results strongly suggest that negative intraspecific effects are stronger than interspecific effects, regardless of whether they are sufficiently strong to maintain long-term stable coexistence (Turnbull et al. 2013).

Using observational data of temporal community change, one can test this prediction by asking whether fitness is higher when species are at low frequency. For example, Harms et al. (2000) assessed the seed-to-seedling transition in small plots in a tropical forest, finding that rarity was associated with a systematic fitness advantage, and thus seedling diversity was greater than seed diversity (see also Green et al. 2014). Other observational studies have employed sophisticated analyses involving the use of field data to parameterize models of each species' population growth as a function of the identity and abundance of potentially competing individuals (e.g., near neighbors in a plant community), as well as potentially confounding covariates. Several studies of this nature (sometimes including experimental data as well) have found intraspecific effects to be more strongly negative than interspecific effects (Adler et al. 2006, Adler et al. 2010, Clark 2010).

Prediction 2b: Species tend to increase in abundance when at very low frequency and when other species "are at their typical abundances" (Siepielski and McPeek 2010).

This is the "invasibility criterion" for coexistence (Chesson 2000b), mentioned earlier. This prediction is more specific than Prediction 2a in that it requires the strength of negative frequency-dependent selection to be sufficiently strong to overcome constant selection.

Methods 2b: Using either direct experimentation, or a model parameterized with observational and/or experimental data, quantify population growth rates for each species when at extreme low abundance and when the rest of the community is at equilibrium.

Results 2b: As mentioned previously, this is a very challenging prediction to test. Of 323 studies on species coexistence reviewed by Siepielski and McPeek (2010), just seven included a test of this prediction. These few studies did indeed generally find support for the invasibility criterion, but most tests were indirect, relying on a model of one sort or another to assess what would happen *if* each species invaded a community that was otherwise at equilibrium (e.g., Adler et al. 2006, Angert et al. 2009). Levine and HilleRisLambers (2009) took things a step further in a series of experiments with annual plants in California. In addition to directly assessing negative frequency-dependent fitness (Fig. 8.6b), they used field-parameterized models to predict each species' fitness in the absence of negative frequency dependence and then in experimental communities in which each species was "forced" to have this fitness. Levine and HilleRisLambers (2009) found that eliminating negative frequency dependence caused a decline in species diversity compared with control communities (i.e., where fitness was determined "naturally"). In sum, there are some compelling examples of local negative frequency-dependent selection being sufficiently strong to overcome constant selection, but there are too few data to assess generality.

Prediction 2c: Among locally co-occurring species, trait values show greater statistical spread or a more regular spacing (along the trait axis) than in random selections of species from the regional pool. This is often referred to as trait "overdispersion" (Weiher and Keddy 1995), and it is the flip side of Prediction 1b.

So far in this section, the critical distinction among the individuals in a community—from the point of view of a particular focal organism—has been whether they are of the same or different species. By definition, selection is based on some phenotypic (i.e., trait) differences between species, and if negative effects of one species on another depend on phenotypic similarity, then this prediction arises in a model of trait-based negative frequency-dependent selection (see Fig. 5.3c).

Method 2c: Observe species composition in a set of sites. For each species, measure relevant traits. Calculate trait variance, dispersion, or a metric of spacing in each plot and in repeated random samples from the regional pool of species (typically the set of species observed across the full set of plots).

The methods here are essentially the same as for Prediction 1b. One can also modify the null model for this prediction to first account for the fact that the local range of trait values is narrower than that in the regional pool. For example, in the null model one can sample trait values only from within the trait range observed locally, which can increase power to test this prediction (Bernard-Verdier et al. 2012).

Results 2c: The last decade has seen a major flurry of studies testing this prediction, and the results have been highly variable. There are certainly compelling examples of trait overdispersion (e.g., Fig. 8.7), but support for this prediction seems to be found considerably less often (i.e., fewer traits in fewer studies) than for Prediction 1b-(i) (i.e., underdispersion) (Vamosi et al. 2009; Kraft and Ackerly 2010; HilleRisLambers et al. 2012; Kraft, Adler, et al. 2015). However, even if the underlying process of trait-based selection is strong, the power to statistically detect overdispersion in natural communities may be much lower than the power to detect underdispersion (Kraft and Ackerly 2010; Vellend et al. 2010, Kraft, Adler, et al. 2015), and few studies first control for underdispersion before testing for overdispersion (Bernard-Verdier et al. 2012). Many studies have tested this prediction assuming that phylogenetic differences between species can stand in as a proxy for relevant ecological

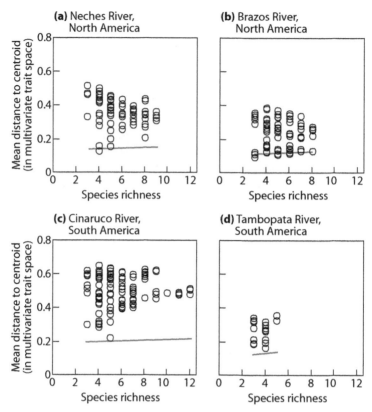

Figure 8.7. Trait overdispersion in fish communities in floodplain rivers of North America (a, b) and South America (c, d). First, 23 morphological traits related to locomotion and feeding were measured on each species and subjected to a principal components analysis (PCA). Trait dispersion in each "mesohabitat" (i.e., for each data point) was expressed as the mean distance of species to the centroid in trait-PCA space; the null-model prediction (gray line) was generated via random samples from the species pool in each region. In all cases, species were overdispersed in trait space relative to the null model (i.e., points tend to lie above the null-model prediction). Data from Montaña et al. (2013).

differences, although there is considerable evidence refuting this assumption (Bennett et al. 2013, Best et al. 2013, Narwani et al. 2013, Godoy et al. 2014, Pigot and Etienne 2015).

FAQ FOR LOW-LEVEL PROCESSES UNDERLYING NEGATIVE
FREQUENCY-DEPENDENT SELECTION

What is the mechanistic basis of negative frequency-dependent selection? Trade-offs among species: The competitive exclusion principle, and Hutchinson's

(1961) declaration of the "paradox of the plankton," spawned a great many studies searching for and analyzing the specific trade-offs among species that might permit their coexistence (Ricklefs and Miller 1999, Tokeshi 1999, Krebs 2009, Siepielski and McPeek 2010, Martin 2014; see also Table 5.1). Examples include relative abilities to compete for different limiting nutrients (Tilman 1982), partitioning of prey resources (Schoener 1974), intransitive competitive networks (Kerr et al. 2002), susceptibility to different pathogens or other enemies (Connell 1970, Janzen 1970), and the use of different microhabitats (MacArthur 1958). For every proposed mechanism, one can find empirical support in the literature for at least some systems, although it is rarely clear that such differences are sufficient to maintain stable species coexistence (Clark 2010, Siepielski and McPeek 2010).

8.3. HYPOTHESIS 3: TEMPORALLY VARIABLE SELECTION IS AN IMPORTANT DETERMINANT OF COMMUNITY STRUCTURE AND DYNAMICS

This hypothesis is not completely independent of the previous two. Specifically, Prediction 1c (based on spatially variable selection) was that an experimental environmental change should lead to community change (over time), and many studies have focused on temporal fluctuations in selection as one pathway to the emergence of long-term negative frequency dependence (e.g., Adler et al. 2006, 2010; Angert et al. 2009). However, as with negative frequency-dependent selection, the potential importance of temporally variable selection goes beyond its connection with spatial patterns or its sufficiency in explaining long-term stable coexistence. In addition to having a major impact on transient community dynamics, it can potentially reduce long-term average fitness differences, thereby reducing the strength of negative frequency-dependent selection needed for coexistence (Huston 2014).

Prediction 3a: Changes in species composition are correlated with environmental conditions across time.

This prediction resembles Prediction 1c, but with an important difference. Prediction 1c involved only a single pulse of environmental change and analysis of community properties before and after such a pulse. Here we are interested in longer-term fluctuations in environmental conditions and whether community composition "tracks" such fluctuations.

Method 3a: In a series of time points, quantify species composition and measure environmental variables thought to underlie selection.

Results 3a: As with composition-environment relationships across space, composition-environment relationships across time seem to be very common. In many community types, including terrestrial plants and vertebrates, marine shell-forming invertebrates, and freshwater phytoplankton, paleoecologists routinely observe correlated changes in community composition and environmental conditions (often climate variables) over hundreds or thousands of years (Davis 1986, Roy et al. 1996, MacDonald et al. 2008, Pandolfi et al. 2011, Jackson and Blois 2015). Direct observations of community change over shorter time frames—years to decades—also reveal frequent correlations between community change and environmental changes such as climate fluctuations (Parmesan 2006) and disturbances of many kinds (Pickett and White 1985).

Predictions 3b: (i) The range or variance in trait values at one point in time is less than that in random selections from the set of species observed during the full time series of data. (ii) Mean trait values are correlated with environmental conditions across time.

Methods 3b: At multiple time points, measure species-level traits, observe community composition, and calculate trait statistics for each time point (mean and range/variance). For Prediction 3b-(i), compare the observed range/variance with expectations based on a null model (see Prediction 1b). For Prediction 3b-(ii), assess correlations between mean traits and environment.

Results 3b: I do not know of any studies conducting explicit trait-based null-model tests of Prediction 3b-(i). However, some studies of tropical forest succession used phylogenetic rather than trait data, along with both space-for-time substitutions and explicitly temporal data, and found that overdispersion rather than underdispersion was the dominant pattern found for trees (Letcher 2010, Norden et al. 2011).

There have been more tests of Prediction 3b-(ii). Studies often involve space-for-time substitutions, showing that community-level average trait values in plant communities vary with time since disturbance (Verheyen et al. 2003, Grime 2006, Shipley et al. 2006), strongly suggesting that mean trait values fluctuate in response to disturbance cycles. Lacourse (2009) applied multivariate analyses to paleoecological data on changing forest communities on the west coast of North America to reveal temporal trait-environment correlations (e.g., greater maximum height during warm periods). Edwards et al. (2013) found that phytoplankton species with different values of several traits (nitrate uptake affinity, light sensitivity of growth, and maximum growth rate) showed predictable responses to fluctuating light and nitrate levels in the English Channel. Moreover, the highly predictable responses of some phytoplankton to temporal environmental changes (e.g., pH, water level, pollution) have allowed researchers to use fossil diatom communities to predict past environmental conditions in many kinds of aquatic habitat (Smol and Stoermer 2010). A similar approach has been used with macroinvertebrates to monitor contemporary environmental change over time in freshwaters (Menezes et al. 2010). Overall, there very often appears to be a correspondence between trait-environment relationships found over space and time, although the number of explicitly temporal studies is dwarfed by the number of spatial studies.

Prediction 3c: Species diversity increases as a function of temporal environmental heterogeneity.

This prediction assumes that the dominant influence of temporal environmental heterogeneity is via its effect on temporally fluctuating selection per se. However, this hypothesis-prediction link is highly prone to the three-box problem (see Fig. 7.2) for at least two reasons. First, if environmental fluctuations cause total community size to fluctuate—as expected if disturbance is involved—then the long-term "effective" community size (Vellend 2004, Orrock and Fletcher Jr. 2005) will be reduced, thus increasing the importance of drift and consequent stochastic extinctions (Adler and Drake 2008). Second, as with spatial heterogeneity, the effects of temporal environmental variance might introduce occasional extreme conditions (e.g., exposure to desiccation), thus involving strong bouts of selection disfavoring one species or another and potentially causing extinctions, as well (Adler and Drake 2008).

This prediction also closely resembles a portion of the intermediate disturbance hypothesis (Grime 1973, Connell 1978), which predicts an increase in species diversity from low to moderate disturbance levels owing largely to reduction in long-term fitness differences between species due to temporally variable selection (Huston 2014). Periodic disturbances represent a very common form of temporal environmental heterogeneity in nature (Pickett and White 1985, Huston 1994). The predicted decline in diversity from moderate to high level of disturbance is thought to occur for reasons other than temporally variable selection per se (drift and constant selection, as described in the previous paragraph) and so is outside the scope of this section of the book. Before proceeding, it is worthwhile noting that disturbance is often so loosely defined—including almost any kind of abrupt change with potentially important ecological impacts (e.g., Krebs 2009)—that the distinction between periodic disturbance and more gradual environmental fluctuations can be rather arbitrary.

Methods 3c: Assess species diversity in a set of sites varying either naturally or experimentally in their degree of temporal environmental heterogeneity.

Results 3c: Many observational and experimental studies have tested for relationships between species diversity and disturbance frequency or intensity, both of which represent temporal environmental heterogeneity. Results are highly variable, with relationships taking many forms—positive, negative, unimodal, and nonsignificant (Mackey and Currie 2001, Hughes et al. 2007). Nonetheless, the results of these meta-analyses indicate that a positive relationship across at least part of the disturbance gradient is fairly common.

Other studies have focused on varying degrees of temporal fluctuations in factors such as light intensity (e.g., Flöder et al. 2002), nutrient supply (Beisner 2001), water supply (Lundholm and Larson 2003), or climate (Adler et al. 2006). For example, Flöder et al. (2002) exposed a community of freshwater phytoplankton to constant light intensity, as well as to different frequencies

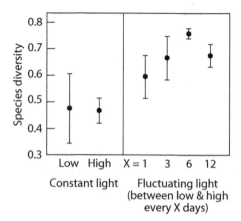

Figure 8.8. Species diversity (Shannon index) of phytoplankton extracted from Lake Biwa, Japan, and grown for 49 days in the lab under constant or fluctuating light intensity. Error bars represent ±1 standard error ($N = 3$ per treatment). Data from Flöder et al. (2002).

of switching between high and low intensity over time. Greater species diversity was observed in the temporally variable environments than in constant environments (Fig. 8.8). In contrast, Lundholm and Larson (2003) found that the species richness of establishing plant seedlings declined with increasing temporal variation in soil moisture (with a constant total water supply). Using field-parameterized models, others have demonstrated the essential role that environmentally driven temporally fluctuating selection can play in allowing stable species coexistence (e.g., Adler et al. 2006, Angert et al. 2009), which may promote species diversity. However, it is important to note that explaining stable coexistence of a given set of species (typically a very small set) is not the same as predicting variation in species diversity among different sites (Huston 2014, Laliberté et al. 2014).

In short, increasing temporal environmental heterogeneity often appears to be associated with increased species diversity, although this result is highly context dependent.

FAQ FOR LOW-LEVEL PROCESSES UNDERLYING TEMPORALLY VARIABLE SELECTION

Which environmental variables and traits underlie selection? Many studies begin with an interest in the specific environmental variables that underlie temporally variable selection, and particular traits that mediate community responses, as evident in the studies described. Tests of these questions involve methods analogous to those used to identify the environmental variables and traits involved in spatially variable selection (see Sec. 8.1).

Under what circumstances does temporally variable selection lead to long-term negative frequency-dependent selection? Temporally fluctuating environmental conditions feature prominently in several low-level models of species coexistence (see Table 5.1), which have inspired empirical studies. For example, in a community of desert annual plants, negative frequency-dependent selection (and therefore stable coexistence) was found to emerge from several attributes of the system, including interspecific differences in demographic responses to fluctuating rainfall (i.e., temporally variable selection), the ability of species to persist during dry periods via dormant seeds in the soil, and strong competition that reduced the population growth of competitive species during favorable time periods (Angert et al. 2009).

8.4. HYPOTHESIS 4: POSITIVE FREQUENCY-DEPENDENT SELECTION IS AN IMPORTANT DETERMINANT OF COMMUNITY STRUCTURE AND DYNAMICS

The study of positive frequency-dependent selection—or, more generally, positive feedbacks—in communities is challenged by the fact that in nature we expect to see it in action only ephemerally, when communities shift from one stable equilibrium state to another (see Fig. 6.6). Thus, empirical studies often employ concepts such as "alternative stable states," "phase shifts," "critical transitions," "tipping points," "priority effects," or "historical contingency," among others (Lewontin 1969, Slatkin 1974, Scheffer et al. 2001, Bever 2003, Suding et al. 2004, Scheffer 2009, Fukami 2015).

Unlike in the case of negative frequency-dependent selection, empirical demonstrations of positive frequency-dependent selection only occasionally involve "simple" scenarios of pairwise species interactions. More commonly, they involve fairly complex feedback loops among different functional groups of organisms on the same or different trophic levels, as well as abiotic environmental variables and disturbance regimes (Scheffer 2009). Nonetheless, the crux of many such examples is shifts among two or three community states along a single axis that essentially represents community trait composition, such as coral versus algae dominance of reefs (Hughes 1994, Mumby et al. 2007), as will be explained further. The term *hysteresis* is often used to characterize situations in which the outcome of a dynamic process depends on initial or historical conditions (see Fig. 8.9c).

Prediction 4a: "Long-term" community dynamics or (quasi)equilibrium community composition is sensitive to initial community composition.

Methods 4a: Experimentally manipulate initial community composition; follow subsequent community dynamics.

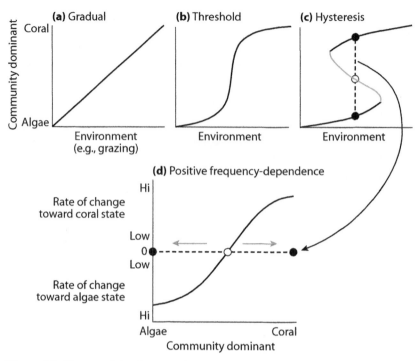

Figure 8.9. Three models of community responses to environmental change (a–c) and the emergence of positive frequency-dependent selection in "intermediate" environments in the hysteresis model. The example of coral- versus algae-dominated reefs is used as an illustrative example (Mumby et al. 2007). Other examples would follow the same logic: savanna versus forest along a rainfall gradient (Hirota et al. 2011), or macrophyte versus plankton dominance along a nutrient gradient (Scheffer et al. 1993). In (a–c), black lines represent stable equilibria, and the gray line unstable equilibria (for a given environmental state). In (c) and (d), filled circles represent stable equilibria, and the open circle represents an unstable equilibrium. Arrows indicate the expected direction of dynamics.

Empirical tests of this prediction are most often reported using the rubric of "priority effects" or "historical contingency," with initial differences in community composition created by experimentally manipulating the order in which species are added to communities (Chase 2003, Fukami 2015). It can be difficult to determine whether a community has reached or is on its way to an equilibrium point; hence the term "quasi-equilibrium" is used in the phrasing of this prediction. However, many experiments of this nature use organisms with short generation times (e.g., bacteria, yeast, plankton, or fruit flies), allowing experiments to run for tens if not hundreds of generations, thus minimizing the

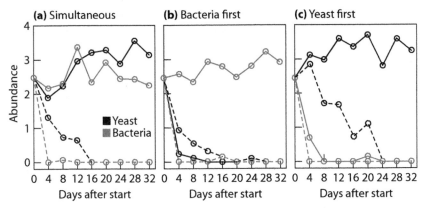

Figure 8.10. Dynamics of two yeast and two bacteria species under different orders of colonization into experimental microcosms of flower nectar. *Metschnikowia reukaufii* (solid black line) and *Gluconobacter* sp. (solid gray line) appear to coexist when introduced simultaneously (a). However, if one or the other is introduced first, it excludes the other (b and c). Two other species (dashed lines), *Starmerella bombicola* (a yeast) and *Asaia* sp. (bacterium) did not persist in these treatments. Abundance was measured in units of $\log_{10}$ (colony-forming units per microliter of nectar + 1); $N = 4$ for each treatment (error bars were very small in all cases and are not shown). Data from Tucker and Fukami (2014).

chance that differences among colonization-order treatments are due only to slow convergence to the same community state.

Results 4a: Results of some studies refute this prediction, finding convergence to the same community composition regardless of initial conditions, while others support the prediction, finding strong dependence of "final" composition on colonization order (reviewed in Chase 2003, Fukami 2015). Many experiments manipulate colonization order from a pool of species that includes multiple trophic levels (primary producers, herbivores, predators, detritovores, etc.), most often in aquatic microcosms. Some studies focus on "horizontal" components of the community, as well. For example, using three alga species in small (250 mL) freshwater microcosms, Drake (1991) found that the abundances of two "poor competitors" were strongly dependent on whether they were introduced before or after the dominant competitor (i.e., whether they started at high or low relative abundance). In larger (40 L) microcosms, the introduction sequence of four primary producers had a large impact on subsequent community dynamics, including the success of various consumers, and their associated feedbacks on the producer community (Drake 1991). Similarly, Tucker and Fukami (2014) observed strong priority effects between yeast and bacteria species inhabiting flower nectar, at least under some conditions (Fig. 8.10).

In interpreting results of these kinds of experiments, one must remain aware that an influence of initial conditions on final community composition is not in and of itself evidence of positive feedbacks, given that the outcome of pure ecological drift is also sensitive to initial species frequencies (e.g., a species starting at a frequency of 0.8 has an 80% chance of drifting to total dominance; see Chap. 6). Some experiments, such as those revealing stronger priority effects in small relative to large communities (Fukami 2004), suggest a role for drift, although many experiments reveal alternative community trajectories more quickly and systematically (i.e., repeatable across replicates) than would be expected based on drift alone (reviewed in Chase 2003, Fukami 2015).

Prediction 4b: Organisms modify their environment in ways that increase the relative fitness of conspecifics.

Method 4b: Experimentally add a species to a site, allow it time to modify its environment, and then assess the relative fitness of conspecifics and heterospecifics in modified and unmodified sites.

Results 4b: This prediction has been especially well developed in studies of feedbacks between plants and their associated soil biota (Bever et al. 1997, Bever 2003, Reynolds et al. 2003). Briefly, plants of various species are added to pots of standardized soil and allowed to grow for a period of time. New plants of various species are then grown in these soils (without the original plant), and components of fitness (e.g., total biomass) are quantified. The most common result is negative feedbacks—that is, plants modify soils to the detriment of conspecifics more than heterospecifics—thus providing support for a variant of Prediction 2a. However, the opposite result—positive feedbacks—is also found for some species in some studies (reviewed in Bever 2003, Bever et al. 2010). Similar studies have been conducted in other community types, as well. For example, Lee (2006) experimentally added coral rubble (i.e., the structure produced by corals without the living organism itself) to patches of an algae-dominated reef and found that the structure enhanced the density of the key herbivore (urchins), thereby depressing algae, and presumably (this was not directly measured) enhancing the fitness of corals themselves. In sum, experiments with pairs of species sometimes show evidence consistent with positive frequency-dependent selection but not as often as they show evidence of constant or negative frequency-dependent selection.

Predictions 4c: (i) Site-to-site variation in community composition (beta diversity) is large despite a lack of initial environmental differences among sites. (ii) For many pairs of species, presence of one species is associated with absence of the other. (iii) Sites under similar environmental conditions support communities of distinctly different types, rather than showing gradual variation in composition.

These three predictions are grouped together given their common core element: sites with similar initial environmental conditions (i.e., no externally imposed spatial variation in selection) should support communities with highly variable composition from one site to the next. The first prediction comes from the priority effects literature and is more generalized, with "large" defined in a comparative sense. Tests of this prediction are based on a priori predictions of factors thought to promote the manifestation of positive feedbacks (Chase 2003). The second prediction, coming from the literature on "assembly rules," is the simplest possible manifestation of a priority effect and is known as a "checkerboard" pattern (Diamond 1975, Weiher and Keddy 2001). The third prediction stems from the literature on multiple stable states, and it involves testing for the clustering of communities in two or more distinct community states (multimodality). The absence or at least rarity of intermediate community states suggests positive feedbacks pushing the community to one of the common (and presumably stable) states, depending on the initial community state (Scheffer and Carpenter 2003). Drift among communities with many species can create differences in community composition (Hubbell 2001), but it does not predict multimodality.

Methods 4c: Assess community composition in one or more groups of sites. If there are multiple groups, the sites within a group should occur in similar environmental conditions. For Prediction 4c-(i), test for a difference in beta diversity among groups of sites. For Prediction 4c-(ii), test for negative associations within pairs of species ("checkerboard" patterns of codistribution), and assess whether the strength and frequency of such associations is greater than in an appropriate null model. For Prediction 4c-(iii), test for multimodality in the distribution of sites (within a group) along one or more axes of community composition.

Results 4c: Prediction 4c-(i) has been most thoroughly tested in a series of experimental and observational studies by Jonathan Chase and colleagues, in which greater beta diversity in freshwater ponds or microcosms was found in communities with a large potential species pool, low disturbance, low fish predation, low stress, and high productivity (Chase 2003, 2007, 2010; Chase et al. 2009). These factors were thought to promote positive feedbacks, and thus the possibility of multiple stable states.

Testing for nonrandom patterns in presence-absence community data, such as checkerboards (Prediction 4c-(ii)), has a controversial history. In a detailed analysis of birds on islands in and around New Guinea, Diamond (1975) reported examples of checkerboard patterns among certain pairs of species, concluding that competition and priority effects precluded those pairs of species from coexisting locally. These conclusions were harshly criticized on the grounds that checkerboard patterns were observed no more often than one would expect even if each species was distributed at random across islands (Connor and Simberloff 1979, Strong et al. 1984). Reanalysis of many such presence-absence data sets suggests that nonrandom patterns, such as checkerboards, are indeed quite common (Gotelli and McCabe 2002). However, such

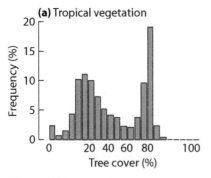
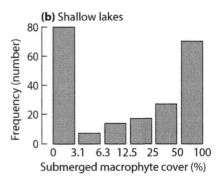

Figure 8.11. Multimodal distributions of community states. (a) Tree cover in Africa, Australia, and South America in areas with moderate rainfall (~1800 mm/year); (b) Cover of macrophytes in 215 lakes in the Lower Rhine floodplain, The Netherlands. The *x*-axis in (a) was arcsine transformed. Data taken from original publications: (a) Hirota et al. (2011) and (b) Van Geest et al. (2003).

analyses do not control for the possibility that such patterns are produced by spatially variable selection—a process of widespread importance (see Hypothesis 1)—so, ultimately, these results constitute very weak evidence of positive frequency-dependent selection.

Community composition often varies gradually along environmental gradients, with sites under similar conditions showing similar composition and no obvious evidence of multimodality (Whittaker 1975). However, a variety of situations have been found in which sharply contrasting communities are found under similar conditions: tree versus grass dominance under moderate rainfall (Hirota et al. 2011; Fig. 8.11a), coral versus algal dominance at moderate levels of grazing (Mumby 2009), and macrophyte- versus phytoplankton-dominated shallow lakes (Scheffer et al. 1993, Fig. 8.11b). The possibility of multiple stable states under relatively dry conditions (e.g., savanna vs. forest) is clearly hinted at in Whittaker's (1975) schematic of how terrestrial biomes map onto climate space (see Fig. 8.1). Wilson and Agnew (1992) and Scheffer (2009) describe many other apparent cases of strongly contrasting community types occurring under similar initial environmental conditions. In these studies, it is important to specify that environmental conditions are similar "initially," given that different subsequent community types can strongly modify local environmental conditions, such as soil properties (Chase 2003). On their own, patterns of this nature also do not allow one to exclude the possibility of a multimodally distributed but unmeasured environmental variable causing the community pattern.

Overall, studies testing these predictions suggest that positive feedbacks are important in nature, albeit under a restricted range of conditions.

Prediction 4d: Communities change rapidly in composition in response to a temporal change in externally imposed selection (i.e., environmental change or disturbance), but these changes do not reverse with a reversal of environmental change of magnitude equivalent to the initial change.

The previous prediction focused on static patterns expected based on the hysteresis model of community responses to environment (see Fig. 8.9c). The current prediction focuses on expected temporal dynamics under the same model.

Methods 4d: Using observation or experiment, assess community responses to a directional change in environmental conditions or disturbance, and the subsequent response to a change in environment or disturbance in the opposite direction.

Results 4d: For the cases described previously for terrestrial vegetation, coral reefs, and shallow lakes, there is some evidence to support this prediction. For example, Dublin et al. (1990) reported long-term observations in the Serengeti ecosystem of East Africa showing drastic reduction in tree cover due to fire but no subsequent increase in tree cover in the absence of fire; herbivores were thought to maintain the treeless state but not to be capable of causing the phase shift on their own. Scheffer et al. (1993) describe several cases of management interventions to reduce nutrient loading in shallow lakes, with no subsequent reversal of the algae-dominated state until nutrient levels were far below those at which the initial phase shift occurred (e.g., Fig. 8.12) or unless fish were also

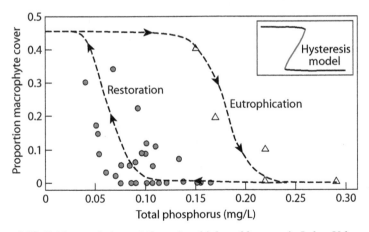

Figure 8.12. Evidence of phase shifts and multiple stable states in Lakes Veluwe and Wolderwijd, The Netherlands. Phosphorus loading caused loss of macrophyte cover in the 1960s (open triangles), but P had to be reduced far lower than the initial level to prompt macrophyte recovery 20 years later (filled gray circles). Arrows indicate the direction of temporal change. The inset shows the hysteresis model for this system, as defined in Figure 8.9c. Data from Meijer (2000).

added to the lake. In some coral reefs, an algae-dominated state was induced by the die-off of a dominant herbivore (urchins), and while algae cover declined to some degree as urchins recovered, the coral-dominated state did not return (Mumby 2009). In a temperate grassland, experimentally imposed nutrient input greatly reduced species richness and promoted dominance by one grass species, but these changes were not reversed even after 20 years without experimental nutrient addition (Isbell et al. 2013). Given the rarity of data available to test this prediction, it is difficult to assess the general applicability of these examples.

FAQ FOR LOW-LEVEL PROCESSES UNDERLYING POSITIVE FREQUENCY-DEPENDENT SELECTION

What specific interactions among species and environmental factors create the feedbacks that underlie positive frequency-dependent selection? As many of the preceding examples indicate, studies testing these predictions often involve a focus on mechanistic details. Supplementing the main examples already introduced, the following descriptions are simplified versions of fairly complex ecological stories.

In the case of terrestrial vegetation, major fires can convert a forest to a savanna or a savanna to grassland, with the greater dominance of grass in the "new" community state itself promoting fire and herbivores that, in turn, prevent a return to the initial community composition (Dublin et al. 1990, Hirota et al. 2011, Staver et al. 2011). In Caribbean coral reefs, a disease-induced die-off of urchins in the 1980s exacerbated other anthropogenic disturbances, permitting establishment of algae dominance, and the coral decline eliminated habitat for herbivorous fish, as well as refuge for urchins, thereby preventing coral recovery (Hughes 1994, Mumby et al. 2007, Mumby 2009). In many shallow lakes, nutrient input caused an increase in productivity, prompting people to remove "nuisance" macrophytes (which were also becoming covered in periphyton), thus removing refuges for phytoplankton-eating zooplankton and permitting wind to bring sediment into suspension, which—along with thriving algae and increased benthic feeders that disturb the sediment—reduced regeneration opportunities and the light levels needed for macrophytes to establish themselves (Scheffer et al. 1993, Scheffer 2009). Other examples involve natural "switches" between different kinds of plant communities driven by positive plant-environment feedbacks (Wilson and Agnew 1992), or various kinds of changes initiated by nonnative species or other anthropogenic impacts (Simberloff and Von Holle 1999, Mack et al. 2001, Suding et al. 2004). Each case study involves a unique set of important low-level processes but a common high-level outcome: the possibility of alternative stable states via positive frequency-dependent selection.

TABLE 8.1. A Summary of the Empirical Support for, and Challenges and Caveats Involved with, Hypotheses and Predictions Based on the Importance of Different Forms of Selection in Ecological Communities

Hypothesis (H) or Prediction (P)		Empirical Support	Challenges and Caveats
H1	Constant and spatially variable selection	Process detected often and fairly easily; likely ubiquitous in nature	
P1a	Composition-environment relationship across space	Thousands of examples support prediction; exceptions extremely rare	Unmeasured environmental variables potentially important
P1b(i)	Small trait range/variance locally	A given study is more likely than not to find some supporting evidence, but not for all traits	Unmeasured traits potentially important
P1b(ii)	Community-level trait-environment relationship across space	A given study is more likely than not to find some supporting evidence, but not for all traits	Unmeasured environmental variables and traits potentially important
P1c	Change in environment leads to change in species composition	Thousands of examples support prediction; exceptions extremely rare	Typically involve few environmental factors, and measure short-term response
P1d	Positive relationship between species diversity and spatial environmental heterogeneity	Most studies support prediction, but a nontrivial minority do not; there are even some examples of negative relationships	Mean environmental conditions and "microfragmentation" of habitat can covary with environmental heterogeneity
H2	Negative frequency-dependent selection	Process can be difficult to detect, but likely operates to some degree in most communities	
P2a	Negative intraspecific effects > interspecific effects	Many supporting studies, as well as many studies finding no such evidence	Often only short-term responses measured; long-term fitness uncertain
P2b	Species increase when rare	Very few clear examples	Strictly defined, prediction is very difficult to test
P2c	Trait overdispersion locally	Many examples, but not found as often as trait under dispersion (P1b(ii))	Lower statistical detection power, even when underlying process is strong
H3	Temporally variable selection	Many examples; process likely of widespread but variable importance, depending on degree of environmental fluctuations	
P3a	Composition-environment relationship across time	Many examples, albeit fewer than for spatial relationships (P1a)	Depending on demographic rates, environmental fluctuations can be too rapid for community-level response (see also P1a)

continued

TABLE 8.1. *Continued*

Hypothesis (H) or Prediction (P)		Empirical Support	Challenges and Caveats
P3b(i)	Small trait range/variance at single time points	Few if any examples, but empirical support seems likely (based on P1b(i) and P3a)	Species pool (used for null model) might strongly resemble local community at one point in time if time series is short (see also P1b(i))
P3b(ii)	Community-level trait environment relationship across time	Many qualitative examples (post hoc interpretations of P3a); fewer quantitative examples	Unmeasured environmental variables and traits potentially important
P3c	Positive relationship between species diversity and temporal environmental heterogeneity	Many examples with disturbance as temporal heterogeneity; fewer for other variables; some studies support prediction, some do not	Temporal environmental heterogeneity can also influence species diversity via drift and constant (on average) selection
H4	**Positive frequency-dependent selection**	Some compelling examples; by definition, not likely to be observed during any one short time period	
P4a	Community dynamics sensitive to initial composition	Many compelling examples; many counterexamples	Sensitivity to initial composition also expected via drift; can be detected even under negative frequency dependence if demographic rates are slow and experiment is short
P4b	Positive intraspecific feedback via environmental modification	Some compelling examples; not as common as those showing negative intraspecific feedbacks	Experiments often short-term; long-term consequences uncertain
P4c(i)	High beta diversity under conditions expected to promote positive feedbacks	Not many studies, but there is support for the prediction in the studies that have been done	Positive feedbacks not measured directly, so risk of defining the process based on the pattern
P4c(ii)	Checkerboard distribution patterns for pairs of species	Found very frequently	Could easily be caused by spatially variable selection
P4c(iii)	Multimodal community composition	Some compelling examples, but not many	Not clear how often studies would have detected this, even if present; could be more common than few examples imply
P4d	Hysteresis	Some compelling examples, but not many	Cases often involve complex interactions, making inferences about process difficult

8.5. SUMMARY OF EMPIRICAL STUDIES OF SELECTION

It is difficult to quantitatively assess how often one might expect to find a particular process in action across the full sweep of ecological communities on earth. However, the massive number of studies carried out does permit at least a qualitative impression of what the answer to this question might look like. My impressions are presented in Table 8.1. Overall, some forms of selection appear to be nearly ubiquitous (e.g., spatially and temporally variable selection), while others are likely to be important only under relatively rare circumstances (e.g., positive frequency-dependent selection). Negative frequency-dependent selection has been of great interest to ecologists given its potential role in stabilizing communities and maintaining diversity, and it seems likely to be of very widespread importance, but its action can be rather difficult to detect.

This chapter has covered a considerable proportion of the domain of community ecology. I hope it has demonstrated how the concept of selection—not often used in ecology outside of its connection with adaptive evolution—can help conceptually organize an otherwise overwhelmingly heterogeneous literature. The same body of empirical evidence could have been presented using alternative headings: coexistence theory, community ordination, resource partitioning, priority effects, niche theory, disturbance, multiple stable states, consequences of environmental heterogeneity, trait-based community ecology, phylogenetic community ecology, competition, biodiversity and ecosystem function, beta diversity, plant-soil feedbacks, paleoecology, and so on. Each of those topics is of immense importance in ecology, but sorting out how one topic relates to the next requires a massive mental effort and many years of study (at least it did for me). I think that the conceptual framework of high-level processes—including selection in its manageable number of forms—can help remedy this situation.

CHAPTER 9

Empirical Evidence:
Ecological Drift and Dispersal

Starting the empirical chapters of this book with the process of selection (see Chap. 8) was not an arbitrary decision. In one form or another, we see selection everywhere, or at least almost everywhere. The outcome of selection is also, for the most part, relatively easy to detect as a statistical departure from a null hypothesis or null model. However, even when statistical signatures of selection are strong and clear (e.g., composition-environment relationships across space and time), a great deal of spatiotemporal variation in community properties typically remains unexplained (Soininen 2014). In other words, demonstrating an important influence of selection on community structure and dynamics does not exclude a potentially important influence of other simultaneous processes, such as drift and dispersal—the subjects of the present chapter. Speciation is considered in Chapter 10.

9.1. HYPOTHESIS 5: ECOLOGICAL DRIFT IS AN IMPORTANT DETERMINANT OF COMMUNITY STRUCTURE AND DYNAMICS

Drift is the result of random sampling during the processes of birth, death, and reproduction. But how can we determine whether something in nature has happened truly at random or whether we simply don't know the nonrandom causes of what has happened? This is a question that has confounded scientists and philosophers of science for centuries (Gigerenzer et al. 1989). The crux of the debate can be restated as questioning whether stochasticity is a fundamental attribute of nature or just a necessary attribute of our models of nature given ignorance of the important deterministic processes (Clark et al. 2007, Clark 2009, Vellend et al. 2014). I proceed from the standpoint that drift is a theoretically viable process for explaining community structure and dynamics, in the sense that demography can involve "practically irreducible probabilities like those in the fall of dice" (Wright 1964) with respect to species identities (see also Chap. 5).

Accepting the theoretical possibility of drift does not negate the difficulty of empirically detecting its importance in ecological communities. Demonstrating that fitness differs between species (i.e., the presence of selection) is more straightforward than demonstrating that it does not (i.e., the absence of selection). As the saying goes, absence of evidence is not evidence of absence. That said, failure to detect seemingly ubiquitous forms of selection (e.g., spatially variable selection), despite due diligence of the researcher, at least suggests that drift can be important. In this section I consider several predictions based on the action of ecological drift; however, I do not include the prediction that relative abundance distributions (see Fig. 2.2) will fit predictions of neutral theory (Hubbell 2001), as this is now widely recognized as an exceedingly weak test in that the same distribution emerges from models including selection (McGill et al. 2007, Rosindell et al. 2011, Clark 2012).

One more important point to be made before delving into the empirical evidence is that we may well expect the conditions under which drift plays an important role to be relatively rare in nature. However, studies of selection can provide some clues as to where to look. For example, as described under Predictions 1a and 1c (see Chap. 8), moving along an environmental gradient often shifts the fitness advantage from one species to another, such that at some point in the middle of the gradient we might expect species to approach functional equivalence, with neither species enjoying a fitness advantage. Likewise, trait-based selection can result in evenly spaced clusters of species in trait space (Prediction 2c), but what determines the relative abundances of those species with very similar traits (Hubbell 2009)? In these two cases, selection might be the dominant process at a large scale or in the entire set of co-occurring species, but this does not obviate the possibility that drift plays an important role in particular times or places, or between particular pairs of species. We can also base predictions on the theoretical result that drift is most important when community size is small.

Predictions 5a: Smaller community size (i.e., relatively few individuals per local community) is associated with (i) lower local diversity, (ii) greater beta diversity, and (iii) weaker composition-environment relationships.

These predictions arise directly from models with or without selection in communities of different size (see Chap. 6). In short, small community size increases the importance of drift, which in turn accelerates local extinction (decreasing local diversity), allows different species to dominate in different places (increasing beta diversity), and sometimes allows species to dominate even where they are not selectively favored by selection (weakening composition-environment relationships). In practice, it is exceedingly rare for community size to be estimated directly, although the size or area of a given habitat unit—presumably a very strong correlate of community size—is frequently studied as a potential determinant of community properties. In many systems, there is no nonarbitrary way of delineating units of habitat, such that studies investigating the influence of area or "ecosystem size" frequently focus on discrete habitat units, such as islands, ponds, forest fragments, or experimental containers.

Methods 5a: In natural or experimental communities of different size, assess community diversity and composition, as well as environmental conditions.

Results 5a: (i) As predicted, species richness almost universally decreases as the area occupied by a given community—and therefore presumably community size—decreases (Fig. 9.1a, c). However, as many authors have argued (MacArthur and Wilson 1967, Connor and McCoy 1979, Williamson 1988, Rosenzweig 1995), several factors other than stochastic drift can contribute to such species-area relationships, most notably spatially variable selection (via environmental heterogeneity). While analyses of some taxa in some places suggest that "area per se" has no detectable influence on species richness after environmental heterogeneity is accounted for, other analyses point to a direct influence of area (Ricklefs and Lovette 1999), possibly via drift.

Results 5a: (ii) I know of no systematic reviews of the question of whether community size is a predictor of beta diversity, but case studies do support Prediction 5a-(ii). For example, for small mammals, Pardini et al. (2005) found that large (>50 ha) forest fragments and areas of continuous tropical forest had similar levels of beta diversity (and species richness), but that beta diversity increased (and richness declined) in medium (10–50 ha) and small (<5 ha) forest fragments (Fig. 9.1 a, b). For understory plants in discrete patches of both primary and postagricultural temperate forest, Vellend (2004) found increasing species richness and decreasing beta diversity as patch size increased (Fig. 9.1 c, d). Similarly, Harrison (1999) found greater plant beta diversity in naturally small fragments of serpentine soils (strongly contrasting with surrounding vegetation) than in equal-sized plots in large areas of such soils.

Results 5a: (iii) I know of only one direct empirical test of the prediction that small community size leads to weak composition-environment relationships. Comparing large (500 m^2) and small (32 m^2) experimentally created patches of grassland, Alexander et al. (2012) found support for this prediction: after 16 years of succession, plant composition-environment relationships were stronger in the larger patches. It is quite possible that more such studies exist (the data are certainly available), but given that I found only one in the literature, I reanalyzed two data sets from my own lab as preliminary tests. First, I divided the sites shown in Fig. 9.1d into two groups (fragments <2.5 ha vs. >2.5 ha), and for each subset I conducted a canonical correspondence analysis

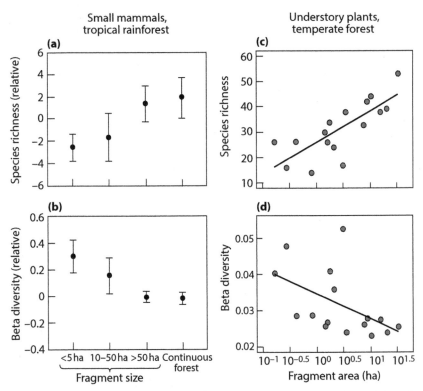

Figure 9.1. Effects of community size on species richness (alpha diversity) and beta diversity. In (a) and (b), mean ± SD of residuals of species richness, and beta diversity from regressions on a composite variable representing forest structure, are plotted for four sizes of forest fragment. Beta diversity was estimated using the pairwise index of Whittaker (1960). Data in (c) and (d) are from the primary forest patches in Vellend (2004), with beta diversity represented by the mean value of a community analogue of F_{ST} (used in population genetics) for a given site paired with all others. Lines in (c) and (d) show best-fit linear least-squares regressions. Data from (a, b) Pardini et al. (2005) and (c, d) Vellend (2004).

(CCA) predicting vegetation composition with a composite axis of soil properties closely correlated with pH (see Vellend (2004) for details on data). The composition-pH relationship was slightly stronger in the larger fragments (% variance accounted for by pH = 23%) than in the smaller fragments (20%). I did the same for 43 fragments of oak savanna habitat on Vancouver Island, Canada (Lilley and Vellend 2009), for which there was a natural gap in the distribution of fragment sizes (26 fragments < 4 ha, 17 fragments > 4 ha). Predictor variables were a composite axis representing climate variation (cool, wet, high-elevation sites vs. warm, dry, low-elevation sites) and the density of roads within 500 m of the fragment. Again, these variables accounted for a greater proportion of compositional variation in the large patches (29%) than in the small patches (21%). It is unknown how general these results are across systems.

Prediction 5b: Differences in community composition are unrelated to differences in environmental conditions between sites.

This prediction is based on the absence of evidence of a widespread form of selection: spatially variable selection. In the absence of selection, and at spatiotemporal scales small enough to negate speciation as an important factor in creating differences among communities, drift is the only process that can create site-to-site compositional variation.

Methods 5b: In a set of study plots in the field, quantify species composition, and measure environmental variables most likely to underlie selection (if present).

Results 5b: As discussed under Prediction 1a (see Chap. 8), species composition almost always shows some degree of correlation with environmental variables. However, for *Enallagma* damselflies in 20 lakes or ponds in the northeastern United States, Siepielski et al. (2010) found no significant correlations between species relative abundances and relevant environmental variables, such as fish density or prey abundance, nor any tendency for community composition to be multimodal (see Prediction 4c(iii)). In a broader set of 40 ponds (including the initial 20), some weak but significant composition-environment relationships were found (Siepielski and McPeek 2013), although overall these results suggest that within certain portions of environmental space, drift is a plausible explanation for spatial variation in community composition (Siepielski and McPeek 2013). Plant ecologists have noted other cases in which considerable site-to-site variation in species composition has no obvious link to environmental variables (Shmida and Wilson 1985, Hubbell and Foster 1986).

Despite some suggestive results, one must always consider the possibility that weak or absent composition-environment relationships result from a failure to measure relevant variables or to adequately assess community composition. In 13/326 data sets included in the meta-analysis of Soininen (2014), <3% of the variation in community composition was explained by environmental variables. However, examination of these studies (e.g., Beisner et al. 2006, Sattler et al. 2010, Hájek et al. 2011) suggests a strong possibility that environmental effects were underestimated rather than absent. In the case of the damselflies studied by Siepielski et al. (2010), independent experimental evidence (see Prediction 5e) bolstered the interpretation of drift as a likely process creating compositional variation among sites.

Prediction 5c: The distribution of trait values among locally co-occurring species is not significantly different from that in random samples from the regional species pool.

This prediction is a third alternative to Predictions 1b and 2c, and in fact, I know of no studies in which there weren't at least some traits that showed nonrandom patterns across sites. This could be due partly to a "file drawer problem" (Csada et al. 1996), although near-ubiquitous causal links between traits, environments, and fitness seem quite probable as well. I include the prediction here for sake of completeness, but I do not pursue it further.

Prediction 5d: The "winner" between two species in competition (i.e., the one that rises to dominance) is unpredictable.

Regardless of the form of selection involved, the action of selection implies that replicate communities with identical initial composition and in identical environmental conditions will follow the same temporal trajectory toward the same stable equilibrium (or possibly a limit cycle). Experimental studies of pairwise species interactions, many of which were described in Chapter 8, often involve small communities (e.g., $J < 100$), in which case drift can, in principle, lead to dominance of one or the other species at random. The only basis for predicting the "winner" would be initial frequency: a species starting at a higher frequency has a higher probability of drifting to dominance (see Chaps. 5 and 6).

Methods 5d: Experimentally create communities of two (or a few) species with identical initial conditions across replicates; follow community dynamics over time.

Results 5d: In a series of classic experiments on flour beetles (*Tribolium confusum* and *T. castaneum*), Thomas Park and colleagues found that the winner in competition switched predictably depending on temperature and humidity conditions (Park 1954, 1962). However, under intermediate temperature-humidity conditions, the outcome was what they called "indeterminate": sometimes *T. confusum* dominated; sometimes *T. castaneum* dominated (Fig. 9.2a). It is difficult to ensure that experimental replicates are truly identical, and some authors suggested that genetic differences among replicate populations underlie a deterministic process driving the seemingly random outcome (Lerner and Dempster 1962). However, even with genetically depauperate source populations, the indeterminate outcome persisted, and moreover, increasing a species' initial frequency also increased its probability of dominating (Mertz et al. 1976), strongly suggesting a prominent role for drift (Fig. 9.2b). Other experiments, many of which focus only on short-term responses (i.e., components of fitness within a single generation), have similarly suggested "competitive equivalence" of particular pairs of species under certain conditions in taxa ranging from phytoplankton (Tilman 1981) to coral reef fish (Munday 2004), and salamanders (Fauth et al. 1990). Importantly, these studies do not exclude the possibility (and indeed some demonstrate it) that selection operates between the same pair of species under different environmental conditions or that selection operates between other species pairs in the same community. They do, however, strongly suggest that the relative abundances of some species, in some communities, under some environmental conditions, fluctuate due to ecological drift.

(a) Proportion of trials "won" by *Tribolium castaneum* in different environments

Temperature	Humidity Low	High
Low	0	29
Medium	13	86
High	10	100

(b) Effect of potential genetic founder effects and initial frequency in intermediate environments

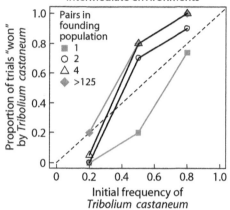

Figure 9.2. Indeterminate competition between flour beetles *Tribolium castaneum* and *T. confusum*. (a) Proportion of trials in which *T. castaneum* dominated under different conditions ($N = 28$–30 per combination; Park [1954]). (b) The probability of "winning" was positively related to initial frequency, and indeterminacy occurred regardless of the potential for genetic founder effects (i.e., the number of pairs in the founding population) ($N = 19$–20 per treatment). Data from Mertz et al. (1976).

Prediction 5e: Intraspecific density and the density of other species have equivalent effects on fitness.

This prediction represents the absence of both constant selection (for at least some species, negative interspecific effects > intraspecific effects) and negative frequency-dependent selection (negative intraspecific effects > interspecific effects; see Prediction 2a).

Methods 5e: In communities with natural or experimentally manipulated variation in the densities of multiple species, test for effects of the density of conspecifics and heterospecifics on fitness.

Results 5e: Chapter 8 provided an overview of the many density-manipulation studies in community ecology, highlighting cases in which results were indicative of either constant or negative frequency-dependent selection. In some cases, fitness or population growth of individual species appears insensitive to the relative densities (i.e., frequencies) of different species, even if total density across species has a negative impact. For example, focusing on two species of the *Enallagma* damselflies mentioned earlier, Siepielski et al. (2010) experimentally manipulated both total density and the frequencies of larvae of the two species. Components of fitness were influenced negatively by total density but were insensitive to species' relative abundances (Fig. 9.3).

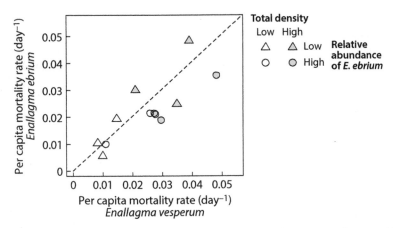

Figure 9.3. In a lab experiment, the per capita mortality rates of *Enallagma ebrium* and *E. vesperum* larvae (damselflies) were strongly influenced by total density but not by relative abundances (i.e., frequencies) of the two species. An influence of frequency would manifest in this graph as circles having high values on the *y*-axis and low values on the *x*-axis, relative to triangles. The dashed line is the 1:1 line. Data from Siepielski et al. (2010).

FAQ FOR LOW-LEVEL PROCESSES UNDERLYING ECOLOGICAL DRIFT

Why do some species seem to be ecological equivalents? Whenever species appear sufficiently similar that selection of any form is weak enough to permit drift, biologists wonder how such species could evolve in the first place. In simple terms, we expect natural selection to lead to species that are either better than or different from their competitors (Rundle and Nosil 2005). In the case of *Enallagma* damselflies, speciation appears to have occurred largely via sexual selection that produced species-specific genitalia incompatible with those of other species, rather than divergence of ecological roles (e.g., resource use) (Turgeon et al. 2005). Hubbell's (2001) neutral theory was inspired by the startling diversity of tropical forest trees, and some models indicate that limited dispersal can prevent strong pairwise species interactions and promote convergence in adaptation to the most common environmental conditions—that is, ecological equivalence (Hubbell 2006). As with the earlier predictions, these studies are the flip side to the many investigations described in Chapter 8 on the ways that species *are* different.

9.2. DISPERSAL

Dispersal is a deceptively simple process. It is simple in the sense that it involves nothing more than the movement of organisms from one site to another, but the consequences of such movement can be quite complex. For selection and drift, one can generate meaningful predictions from a model of community dynamics in a single site, without any reference to dispersal except to assume that there is no immigration into the site. In contrast, dispersal must by definition involve multiple sites, and one cannot build a community-level model involving dispersal without simultaneously specifying parameters that determine local selection or drift. Since the consequences of dispersal can depend on such details, a complex array of outcomes is possible (Leibold et al. 2004, Holyoak et al. 2005, Haegeman and Loreau 2014). Therefore, I do not begin this section with a generic "dispersal is important" hypothesis but, rather, go straight to more specific hypotheses (see the next section).

Dispersal is also interesting because it can represent both a high-level process and a low-level process underlying selection at the regional scale (see also Chap. 5). Considering dispersal as a high-level process, one can essentially ask how local and regional community properties depend on the degree to which individuals—of all species—disperse among sites (e.g, Mouquet and Loreau 2003, Cadotte 2006a). From the point of view of a low-level process, species can and often do vary in their dispersal ability, and dispersal ability might correlate with other traits that collectively underlie various forms of selection (Lowe and McPeek 2014).

Given these considerations, a comprehensive treatment of all possible consequences of dispersal is beyond the scope of this book. I have structured this section as follows. First, I state one general hypothesis (with attendant predictions) about the consequences of dispersal that depend minimally on the local details of selection or drift, assuming only that at least some dispersing propagules or individuals are able to establish in the sites they arrive at (Sec. 9.2.1). I then present one example of a hypothesis and prediction based on an explicit interaction between dispersal and selection (Sec. 9.2.2). Finally, I address how dispersal can represent a low-level process underlying selection (Sec. 9.2.3).

9.2.1. Dispersal as a High-Level Process. Hypothesis 6.1: Dispersal Increases Species' Occupancy of Sites and Leads to a More Even Spread of Individuals across Sites

As mentioned previously, this hypothesis makes the minimal assumption that at least some dispersing individuals are able to establish themselves in the sites to which they disperse. The predictions flow out of population-genetic models both with and without selection (Hartl and Clark 1997), as well as their ecological analogues (MacArthur and Wilson 1967, Shmida and Wilson 1985, Hubbell 2001; see also Chap. 6), at least over a range of dispersal rates from low to moderate (Mouquet and Loreau 2003). One potential consequence of very high dispersal in a particular model of selection is addressed under Hypothesis 6.2.

Prediction 6.1a: Local species diversity increases as a function of increasing dispersal.

Methods 6.1a: In sites that vary naturally or via experimental manipulation in the rate of immigration (i.e., incoming dispersal) or in sets of sites (meta-communities) that vary in their degree of interconnectivity, quantify species diversity.

Dispersal is notoriously difficult to quantify accurately (Nathan 2001), so tests of this prediction and the next often rely on proxies for dispersal, such as the degree of isolation of a site from (or connectivity to) potential sources of immigrants. This is the case for most, albeit not all (e.g., Simonis and Ellis 2013), observational studies. It is also the case for many experimental studies, in which dispersal is permitted to varying degrees by connecting experimental units (e.g., small containers or plots) with tubes or corridors, rather than directly transferring individuals among experimental units, although some studies do manipulate dispersal directly.

Results 6.1a: Multiple lines of evidence support this prediction. First, the theory of island biogeography (MacArthur and Wilson 1967) was inspired, in part, by the oft-noted tendency for isolated oceanic islands to have depauperate floras and faunas relative to less isolated islands or equivalent areas of mainland (Whittaker and Fernandez-Palacios 2007). For example, in an analysis of 346 oceanic islands worldwide, Kalmar and Currie (2006) found that distance to the nearest continent was one of three key variables (along with area and climate) needed to predict bird species richness (Fig. 9.4a). At relatively large scales (km^2), increased dispersal among previously isolated sites, either due to geologic events (e.g., the formation of the isthmus of Panama ~3 million years ago) or in more recent times to human-mediated transport, has also often led to increased species diversity (Vermeij 1991, Sax and Gaines 2003, Sax et al. 2007, Helmus et al. 2014, Pinto-Sánchez et al. 2014).

Smaller-scale studies involving seed addition to plant communities or increased connectivity among microcosms (most often of aquatic invertebrates) typically result in an increase in local species diversity (Cadotte 2006a, Myers and Harms 2009; Fig. 9.4b, c), although there are exceptions in which dispersal had no effect (Warren 1996, Shurin 2000, Forbes and Chase 2002). A repeated experiment examining microarthropod communities in small moss patches connected (or not) by narrow corridors found a positive effect of corridors on local diversity in some years (Gilbert et al. 1998; Gonzalez and Chaneton 2002; Fig. 9.4c) but not others (Hoyle and Gilbert 2004). In one of the very few large-scale field experiments, Damschen et al. (2006) found increased local plant species diversity in 1-ha patches of open habitat when connected by a 150 m × 25 m corridor, relative to isolated controls (Fig. 9.4c).

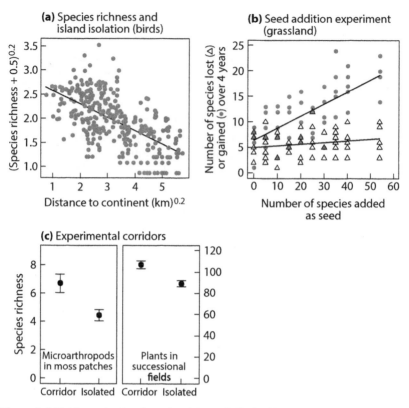

Figure 9.4. Evidence that local species richness is limited by dispersal. (a) For oceanic islands across the globe, isolation was one of three important predictors (along with area and climate) of bird species richness. (b) Adding seeds of species not already present in grassland plots (1m²) resulted in a net gain of species after 4 years, with the gain in species increasing as a function of the number of species added. (c) Creating experimental corridors between small patches (20 cm²) of moss on rocks or between 1-ha patches of successional field increased species richness relative to controls for microarthropods and vascular plants, respectively. Lines in (a) and (b) show best-fit linear least-squares regressions. Data from (a) Kalmar and Currie (2006), (b) Tilman (1997), and (c, d) Gonzalez and Chaneton (2002) and Damschen et al. (2006).

Prediction 6.1b: Compositional dissimilarity among sites (i.e., beta diversity) decreases as a function of dispersal.

Methods 6.1b: Same as the methods for 6.1a, but with quantification of beta diversity.

This prediction has been tested in two related ways. First, assuming that dispersal between pairs of sites declines with geographic distance, one can test for a negative relationship between pairwise beta diversity and distance, after controlling for environmental similarity, which may be confounded with distance (see Prediction 1a, Chap. 8). An important methodological consideration here is how "space" is represented in statistical models. One family of multivariate models uses the x- and y-coordinates of sites to essentially identify a function of x and y that best predicts axes of variation in community composition. These functions can take the form of polynomials (e.g., composition $= f(x, x^2, x^3, y, y^2, y^3)$) or, more recently, combinations of many sine waves of different periodicity to flexibly "find" potentially complex spatial structure in the data at multiple scales (Borcard et al. 1992, Borcard and Legendre 2002). While these methods are excellent ways of exploring spatial structure in multivariate data, there is no clear link between the spatial "signal" that comes out of such analyses and theoretical predictions based on dispersal (Gilbert and Bennett 2010, Jacobson and Peres-Neto 2010), despite many studies claiming just such an association (Gilbert and Lechowicz 2004, Cottenie 2005, Legendre et al. 2005, Beisner et al. 2006, Logue et al. 2011). If, for example, community composition is predicted by an east-west sine wave with a period of 100 m, this implies that two plots 100 m apart have systematically more similar species composition than two sites 50 m apart; drift and limited dispersal is not expected to create such a pattern. Limited dispersal does, however, predict a monotonic decline in compositional similarity with distance (Hubbell 2001), which can be tested directly (Tuomisto and Ruokolainen 2006).

The second way in which this prediction has been tested is by comparing replicate sets of local communities (i.e., whole metacommunities) with different degrees of natural or experimentally imposed connectivity or dispersal among them.

Results 6.1b: With respect to environmental versus spatial determinants of species composition, the only meta-analysis of which I am aware (Cottenie 2005) characterized space using third-order polynomials of the x-y coordinates and thus doesn't speak directly to the prediction stated here. Similarly, hundreds of data sets show a decline in pairwise beta diversity with geographic distance (Nekola and White 1999, Soininen et al. 2007), but without controlling for environmental differences, dispersal is only one candidate explanation. Nonetheless, many individual studies control for environmental differences using the "distance-based" statistical approach (Tuomisto and Ruokolainen 2006), some of which find that, indeed, compositional similarity declines as a function of distance (e.g., Cody 1993, Tuomisto et al. 2003, Barber and Marquis 2010; Fig. 9.5a). Inferring that such patterns are created by spatially limited dispersal can be bolstered by evidence that dissimilarity-distance relationships are stronger for poorer dispersers (e.g., fish vs. plankton; Shurin et al. 2009).

Some studies comparing replicate natural or experimental metacommunities also support this prediction. Most of the studies in the meta-analysis of Cadotte (2006a) did not report beta diversity, but in various types of aquatic microcosm,

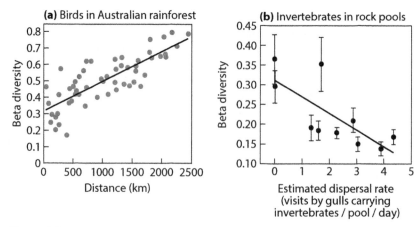

Figure 9.5. Beta diversity as a function of distance and dispersal. (a) For eleven 5-ha rainforest plots (55 pairwise combinations) in eastern Australia, beta diversity (quantified using a presence-absence index) was not correlated with environmental differences (vegetation structure) but was strongly related with geographic distance. (b) Simonis and Ellis (2013) estimated dispersal of invertebrates by seagulls in 10 rock-pool metacommunities and found beta diversity (multivariate dispersion based on Sorensen's index) to decline as a function of estimated dispersal. Points in (b) represent the mean ± SE. Lines show best-fit linear least-squares regressions. Data from (a) Cody (1993) and (b) Simonis and Ellis (2013).

rock pools, or ponds, beta diversity was found to be reduced with increased dispersal or connectivity (Chase 2003, Kneitel and Miller 2003, Cadotte 2006b, Pedruski and Arnott 2011, Simonis and Ellis 2013, Fig. 9.5b), although this result is not universal (e.g., Howeth and Leibold 2010).

> 9.2.2. Dispersal Interacting with Selection. Hypothesis 6.2: Very High Dispersal Permits Constant Selection Regionally to "Swamp" Spatially Variable Selection among Localities, Thereby Reducing Local Diversity

This hypothesis represents one example of how dispersal might interact with selection, which I have chosen to highlight as it emerges from a model that assumes forms of selection for which empirical evidence is especially strong, and because the model's main prediction has itself been empirically tested. Specifically, in a model described as characterizing "mass effects" (see Table 5.1), Mouquet and Loreau (2003) assumed spatially variable selection with a different species favored in each of 20 local sites. As for Prediction 6.1a, when dispersal increases from very low to moderate levels, local diversity increases in this model. However, when the magnitude of local advantage varies among sites (and therefore among species)—a likely scenario— increasing dispersal above a threshold causes local diversity to decline. The threshold was ~30% of individuals dispersing out of one patch and equally distributed among the others in a given time step. In effect, when dispersal is very high, the large local selective advantage enjoyed by some species is extended across the entire metacommunity via a constant influx of propagules to all sites (see also Fig. 6.9).

Prediction 6.2: Local diversity shows a hump-shaped, unimodal relationship with dispersal.

Methods 6.2: Same as for Prediction 6.1a.

Results 6.2: Some studies in aquatic microcosms or habitats support this prediction (Cadotte 2006b, Vanschoenwinkel et al. 2013), and comparisons across studies with animals revealed a significant but weak unimodal diversity-dispersal relationship (Cadotte 2006a). No such relationship was found in comparisons across studies of plants (Cadotte 2006a). Most of the studies mentioned here found monotonic increases or no response of diversity to increasing dispersal, rather than unimodal response, and even those studies finding unimodal relationships did so only under some conditions (Cadotte 2006b, Vanschoenwinkel et al. 2013). That said, studies with only two levels of dispersal (e.g., sites connected or not) would not be able to detect a unimodal relationship, and the finding of a positive diversity-dispersal relationship is not incompatible with the unimodal prediction if dispersal never exceeds the relevant threshold (Cadotte 2006a). However, one must wonder whether the theoretical "tipping point" between positive and negative effects of dispersal (~30%) is so high that pairs of sites recognizable in nature as separate "patches" (e.g., ponds, forest fragments, or rocky outcrops) very often exchange such a high proportion of individuals. With microcosms, one can easily manipulate dispersal to such high levels, but the relevance to natural systems is questionable.

9.2.3. Dispersal as a Low-Level Process. Hypothesis 6.3: Dispersal or Colonization Ability Is a Component of Fitness That Varies across Species and Thus Represents a Trait That Can Be the Target of (a) Spatially Variable Selection or (b) Negative Frequency-Dependent Selection

These forms of selection might arise as follows: (a) If sites vary over the long-term in isolation from other sites, dispersal traits can be the target of spatially variable selection. (b) Dispersal or colonization ability might correlate negatively with competitive ability, thus representing a trade-off axis generating negative frequency-dependent selection at the regional scale (see Fig. 6.10). The latter idea is the basis of the "colonization-competition hypothesis" (Levins and Culver 1971, Yu and Wilson 2001, Cadotte et al. 2006).

Prediction 6.3a: Species composition and dispersal traits vary as a function of site isolation.

Methods 6.3a: Quantify site isolation and other potentially confounding environmental variables, species composition, and dispersal traits.

Results 6.3a: As mentioned earlier, isolated islands are often relatively species poor, and in addition, the species that colonize such islands are not a random sample of the mainland species pools. For example, the only terrestrial mammals native to some isolated oceanic islands (e.g., Hawaii, New Zealand) are those that can fly (i.e., bats), and most forest plants on isolated islands in the Pacific Ocean required transportation by birds for colonization (Carlquist 1967, 1974). Similarly, on islands in a reservoir in the southeastern United States, Kadmon and Pulliam (1993) found that distance to the shore predicted plant species composition, and Kadmon (1995) showed further that poor dispersal was related to exclusion from distant islands (Fig. 9.6a). In some studies of forest fragments in largely agricultural landscapes, average dispersal ability of plant species also appears to increase as a function of site isolation (Dzwonko 1993, Jacquemyn et al. 2001, Flinn and Vellend 2005). Likewise, the effect of isolation on site occupancy appears to decline as a function of dispersal ability within different groups of vertebrates (Prugh et al. 2008; Fig. 9.6b). In sum, isolation can act as an agent of selection.

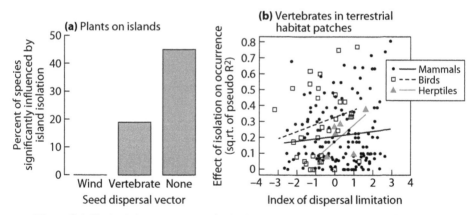

Figure 9.6. Site isolation as an agent of selection. (a) On islands (<1 ha) in a freshwater reservoir, plants with a capacity for long-distance dispersal (via wind or vertebrate ingestion) were less likely to have distributions influenced by island isolation than species with no known capacity for long-distance dispersal. (b) In a meta-analysis, vertebrates showed highly variable sensitivity to site isolation (percent variance in occupancy explained by isolation), and sensitivity increased with dispersal limitation (the log ratio of the maximum distance between habitat patches and the maximum recorded dispersal distance for the species). Lines in (b) show best-fit linear least-squares regressions. Data from (a) Kadmon (1995) and (b) Prugh et al. (2008).

Prediction 6.3b: Traits that promote dispersal or colonization ability are neg-atively correlated with other fitness components (e.g., competitive ability) across co-occurring species.

Methods 6.3b: In a set of co-occurring species, measure traits related to dis-persal/colonization ability and competitive ability.

Results 6.3b: The evolution of organism life histories is subject to funda-mental constraints and has resulted in widespread correlations among traits, suggesting trade-offs (Roff 2002). In relation to the colonization-competition hypothesis, plants frequently show a trade-off between the number and the size of seeds (Harper 1977, Turnbull et al. 1999, Leishman 2001; Fig. 9.7a). Pro-ducing many small seeds may confer a selective advantage in colonizing dis-turbed sites, while seedlings from large seeds potentially outcompete seedlings produced by smaller seeds, giving them a "head start" in terms of resources (Rees and Westoby 1997). Empirical data from competition experiments in-deed indicate that larger-seeded species have a competitive advantage over smaller-seeded species (Turnbull et al. 1999, Freckleton and Watkinson 2001, Levine and Rees 2002). Similarly, some studies also have shown that high seed production and/or dispersal distances of seeds confer an advantage in coloniz-ing disturbed microsites (Platt 1975, Yeaton and Bond 1991). Other studies have found no clear fitness advantage of large-seeded species in competition, or of small-seeded species in colonization (Leishman 2001, Jakobsson and

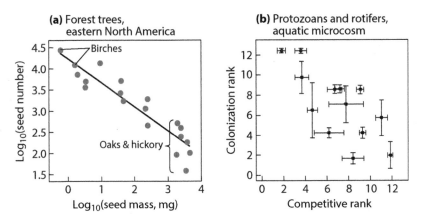

Figure 9.7. Trade-offs hypothesized to underlie regional-scale negative frequency-dependent selection. (a) A trade-off between seed size and number in eastern North American trees; (b) a negative relationship between rank orders for colonization ability (rate of expansion among a set of five connected patches) and competition ability (from pairwise experiments) in aquatic microcosms. The line in (a) shows the best-fit linear least-squares regression. Data from (a) Greene and Johnson (1994) and (b) Cadotte et al. (2006).

Eriksson 2003). Similar studies with animals are comparatively rare, although in an aquatic microcosm assemblage of 13 protozoans and rotifers, Cadotte et al. (2006) found a strong negative relationship between experimentally measured colonization and competition abilities (Fig. 9.7b). In sum, traits thought to be related to colonization and competition ability are often, but not always, negatively correlated.

Despite clear evidence of trade-offs in some cases, reviews of this topic all agree that empirical evidence linking the trade-off to stable species coexistence is very thin (Amarasekare 2003, Levine and Murrell 2003, Kneitel and Chase 2004), notwithstanding a few compelling candidate examples (Platt 1975, Rodríguez et al. 2007, Yawata et al. 2014). Many studies draw very indirect inferences about coexistence based on the existence of the trade-off itself, but parameterizing community models with such data does not indicate that such trade-offs are sufficient to explain coexistence (Levine and Rees 2002, Clark et al. 2004). Nonetheless, the clear fitness consequences of dispersal in many cases strongly indicate that dispersal traits can provide the basis for selection of various forms (Lowe and McPeek 2014).

FAQ FOR LOW-LEVEL PROCESSES UNDERLYING THE EFFECTS OF DISPERSAL

What is the distribution and direction of dispersal distances? Simply quantifying how far, in what direction, and into which habitats dispersers move is a major empirical challenge and the focus of much research (Nathan 2001, Clobert et al. 2012). In many species—plants especially—a very small fraction of dispersers moves much farther than the average dispersal distance, and because such "long-distance" dispersal has important consequences for many phenomena (including diversity increase via colonization of unoccupied sites), it has been the focus of many theoretical and empirical studies (Vellend et al. 2003, Nathan 2006). In many systems, the physical environment constrains the directionality of dispersal (e.g., in river networks), and the consequences of the resulting "topology" of dispersal have also been of considerable interest (Carrara et al. 2012).

How do real communities map onto metacommunity frameworks? Elaboration of four metacommunity "frameworks" (Leibold et al. 2004)—neutral, species sorting, mass effects, and patch dynamics—has inspired much research over the past decade on the community-level consequences of dispersal. In the terminology of the theory of ecological communities, the neutral framework involves drift and dispersal, species sorting represents very strong spatially variable selection, mass effects involve spatially variable selection that is not strong enough to prevent immigrants from establishing sink populations, and patch dynamics involves local extinction (via selection or drift) and (re)colonization via dispersal. These four frameworks essentially represent a classification of spatial community *models* rather than processes, patterns, or dynamics, and not surprisingly, most natural systems include elements reflected in multiple frameworks (Logue et al. 2011). For reasons described earlier (see Table 5.1), I think we can conceptualize community dynamics in a more straightforward and comprehensive way based on four high-level processes than by using these frameworks, but the conceptual mapping from one to the other is fairly simple.

9.3. SUMMARY OF EMPIRICAL STUDIES OF DRIFT AND DISPERSAL

Ecological studies explicitly testing for effects of drift and dispersal are decidedly fewer in number than studies of selection. Thus, general conclusions are necessarily accompanied by greater uncertainty. My impressions—largely qualitative—of the degree of support for each of the hypotheses in this chapter are presented in Table 9.1.

For drift, finite community size ought to make some random drift inevitable, but the ubiquity of selection would appear to marginalize the role of drift. The most compelling empirical examples suggest that drift plays an important role under special circumstances. For example, if speciation has occurred largely via sexual selection with little ecological differentiation—as appears to the be the case for eastern North American damselflies (Turgeon et al. 2005)—the resulting set of co-occurring species might be quite prone to ecological drift (Siepielski et al. 2010). If species are differentiated with respect to the environmental conditions under which each performs best, drift ought to play an important role in "intermediate" environments where selection is weakest, as in the case of *Tribolium* beetles under different temperature and humidity conditions (Park 1954, 1962; Mertz et al. 1976). Patterns of alpha and beta diversity in relation to island or patch size are often consistent with some role for drift in community dynamics, but exceedingly few studies have experimentally addressed community size per se (i.e., without patch size being confounded with other variables). In short, it seems likely that selection very often relegates drift to very minor importance but that drift can be important under some circumstances, in addition to contributing to some widespread community patterns (e.g., species-area relationships).

Empirical evidence is unambiguous in establishing dispersal as an important contributor to community dynamics and structure. However, even theoretical models point to innumerable ways in which dispersal can interact with selection (Mouquet and Loreau 2003, Leibold et al. 2004, Holyoak et al. 2005, Haegeman and Loreau 2014), such that a concise treatment of hypotheses, predictions, and empirical evidence is quite difficult. With respect to the hypotheses and predictions presented in this chapter, local species diversity appears to be frequently limited by dispersal: increasing immigration typically increases diversity. Dispersal also appears quite often to homogenize composition across sites, although there are fewer data with which to assess the generality of this

TABLE 9.1. A Summary of the Empirical Support for, and Challenges and Caveats Involved with, Hypotheses and Predictions Based on the Importance of Drift and Dispersal in Ecological Communities

Hypothesis (H) or Prediction (P)		Empirical Support	Challenges and Caveats
H5	Ecological drift	Many compelling examples; alternative interpretations often possible	
P5a(i)	Small community size = low α diversity	Smaller communities almost always have fewer species	Community size confounded with heterogeneity, edges (and likely other factors), even in experiments
P5a(ii)	Small community size = high β diversity	Smaller communities often show higher β diversity	Community size confounded with other variables (see above)
P5a(iii)	Small community size = weak composition-environment relationship	Very few tests; some suggestive preliminary results in support	Community size confounded with other variables (see above)
P5b	Lack of composition-environment relationship	Very few (but not zero) compelling examples	Unmeasured environmental variables potentially important
P5c	Local trait distribution random with respect to regional pool	None, as far as I know	Unmeasured traits potentially important (if there were supportive cases)
P5d	Unpredictable outcome of competition	Some compelling examples; many studies suggestive of fitness equivalence under particular conditions	Homogenous conditions and small community size in lab may not often extrapolate to the field
P5e	Equivalent intra- and interspecific density dependence	Few (but not zero) compelling examples	Often only some fitness components measured; other fitness components might show different results
H6.1	Dispersal spreads species across sites	Widespread supportive evidence	
P6.1a	Increased dispersal increases α diversity	Strong support via several lines of evidence; some exceptions	Dispersal rarely measured directly; surrogates possibly confounded with other variables
P6.1b	Increased dispersal decreases β diversity	Many compelling examples; not a huge number of tests	Effects of "space" often incorrectly interpreted in terms of dispersal (see also previous challenge/caveat)

TABLE 9.1. *Continued*

Hypothesis (H) or Prediction (P)		Empirical Support	Challenges and Caveats
H6.2	**Very high dispersal enhances region-wide constant selection**	Few good tests	
P6.2	Unimodal diversity-dispersal relationship	Some compelling examples; more cases of monotonic diversity-dispersal relationship	Theoretical threshold between positive and negative effects of dispersal likely very high relative to dispersal among patches in nature
H6.3	**Dispersal is a trait that underlies selection**	Dispersal differences between species often a source of fitness differences	
P6.3a	Dispersal-trait composition varies with site isolation	Some compelling examples; other examples find no relationship	Results often reported at species level; community-level consequences inevitable, but not reported directly
P6.3b	Colonization-competition trade-off (potentially underlies regional negative frequency-dependent selection)	Suggestive relationships between traits common, but consequences for selection usually not clear	Intuitive appeal of colonization-competition logic greater than solid empirical evidence

result. Specific models of dispersal-selection interactions (e.g., Mouquet and Loreau 2003) have been tested relatively few times, with the evidence quite variable from study to study. Finally, dispersal can clearly be involved as a low-level process underlying selection. The colonization-competition hypothesis for species coexistence (Levins and Culver 1971) has endured given widespread observations of potentially relevant trade-offs (e.g., seed size vs. seed number), but the actual community-level consequences of such trait correlations remain poorly understood.

CHAPTER 10

Empirical Evidence:
Speciation and Species Pools

10.1. SPECIATION, SPECIES POOLS, AND SCALE

Many if not most empirical studies of ecological communities at the local scale take as a "given" the regional species pool, the composition and diversity of which need not be explained to study the influence of selection, drift, and dispersal on local community patterns and dynamics. This is a reasonable assumption, notwithstanding the occasional rapid speciation event that can occur at a local scale (e.g., via polyploidy). However, when the level of analysis is an entire region (e.g., the tropics vs. the temperate zone), we can make no such assumption. In this case, the regional species pool effectively becomes the community of interest, and we must incorporate speciation to make a logically complete set of candidate processes underlying observed patterns. At this larger scale, species are added to the community not only by immigration but by speciation, as well (Ricklefs 1987, McKinney and Drake 1998, Magurran and May 1999, Losos and Parent 2009). To reiterate a point made in Chapter 5 (see also Fig. 4.2), I do not include extinction as a separate process in my framework but rather as one possible outcome of selection and/or drift, even if many studies do quantify extinction rather than its causes.

As argued in Chapters 3 and 5, speciation can also be a key component of explanations for some community-level patterns even at very small scales. This is especially true for patterns of species diversity: under certain environmental conditions (e.g., high productivity), local diversity may be limited by the number of species in the regional pool that have evolved to thrive under those conditions, rather than local selection or drift. While *composition*-environment relationships are indicative of spatially variable selection (see Chap. 8), *diversity*-environment relationships on their own have very little to say about the processes that created them. Perhaps such patterns arise because the form and strength of selection varies according to environmental conditions, or perhaps they arise because the regional species pool contains different numbers of species adapted to different abiotic environmental conditions (Ricklefs 1987, Taylor et al. 1990, Zobel 1997). In other words, the explanation for the local-scale pattern might reside in a regional-scale pattern, the explanation of which requires that we consider speciation alongside selection, drift, and dispersal.

Given the preceding arguments, this chapter largely concerns empirical studies of community dynamics and patterns at fairly large scales, or those testing for links between large- and small-scale patterns and processes. In fact, I think that the four-process framework of this book provides an effective conceptual bridge between community ecology done at large scales ("macroecology") and community ecology done at small scales (what we might call "microecology") (see Box 10.1). Hypotheses and predictions presented in this chapter also focus largely on explaining patterns of species diversity, rather than composition. Speciation certainly does influence composition, for example, by producing different species on different land masses even under very similar environmental conditions, thereby increasing large-scale compositional variation (i.e., beta diversity). However, this has been known for centuries, dubbed "Buffon's law" after an eighteenth-century naturalist (Lomolino et al. 2010), and so is not pursued further here.

10.2. IN EMPIRICAL PRACTICE, SPECIATION = SPECIATION + PERSISTENCE

In studies focusing on large-scale community patterns, selection and drift are necessarily treated in a relatively coarse way, although their collective influence on setting regional limits to diversity features prominently in several hypotheses (Ricklefs and Schluter 1993a, Gaston and Blackburn 2000, Rabosky 2013). Instead of a reference to specific forms or low-level underpinnings of selection or drift, the conceptual starting point in macroecology is some version of the following equation (see also Chap. 4):

$$S_t = S_0 + (\text{speciation} - \text{extinction} + \text{immigration}) \times \text{time}$$

In this equation, S_t is the number of species at time t, and S_0 is the number of species at some past time. Speciation (often described as "origination"), extinction, and immigration are all rates, and selection and drift enter the discussion as potential moderators of these rates. Time is not always shown explicitly in this equation, but I've included it here given that it features prominently in some empirical studies.

BOX 10.1.
FOUR HIGH-LEVEL PROCESSES AS A
BRIDGE BETWEEN MICRO- AND MACROECOLOGY

Despite a major wave of studies over the past 20 years explicitly integrating community-level studies across spatial scales (Ricklefs and Schluter 1993a, Leibold et al. 2004, Wiens and Donoghue 2004, Logue et al. 2011), there is a persistent tendency to employ different conceptual frameworks (implicitly or explicitly) in small-scale studies (e.g., vegetation plots in a forest, or ponds in one region) and in large-scale studies (e.g., across continents). At the small scale, we talk about competition, environmental stress, disturbance, dispersal, and so on, while at the large-scale we discuss speciation, extinction, dispersal, and climatic constraints. Dispersal is obviously one important link across scales (Leibold et al. 2004), and I believe that the framework presented in this book can provide an effective and more comprehensive bridge across scales. Specifically, as described in earlier chapters, most invocations of "local processes" (competition, disturbance, etc.) essentially involve selection in its various forms, and occasionally drift as well. At regional scales, researchers often refer to issues like "ecological limits to diversification," rather than the more specific set of factors invoked at the local scale, but essentially they are talking about the same thing: selection (and possibly drift). Finally, while dispersal and speciation might be the most important sources of new species at very small and very large scales, respectively (Fig. B.10.1), both are of potential importance across the full range of scales, with a continuous rather than abrupt shift in relative importance (Rosenzweig

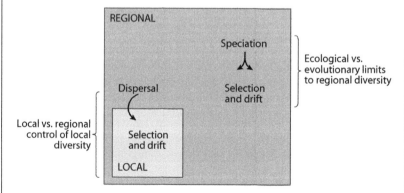

Figure B.10.1. An application of the four-process framework emphasized in this book to illustrate the conceptual parallel between debates concerning the dominant controls over local and regional species diversity.

(Box 10.1 continued)

1995, Gillespie 2004, Rosindell and Phillimore 2011, Wagner et al. 2014). Using the conceptual framework of this book, two major debates concerning species diversity— local versus regional determinants of small-scale diversity (Ricklefs 1987), and evolutionary versus ecological limits to regional diversity (Wiens 2011, Rabosky 2013)—can be understood in very similar terms, distinguished only by the assumed dominant source of species input (Fig. B.10.1). In both cases the key question is whether diversity is limited by the rate of input—dispersal in one case, speciation in the other—or by constraints imposed by selection and drift.

In one sense, speciation influences communities much like dispersal does in that it is a source of new species to the community. However, the nature of the data used to quantify speciation necessitates an important difference. In principle, dispersal itself (i.e., the movement of individuals between localities) can be estimated independently of colonization (i.e., the successful establishment of dispersers in new places), even if this is done indirectly in most cases. In contrast, whether fossil data or molecular phylogenies (the two most common ways to estimate speciation rates) are used, short-lived or incipient species will almost never be observed (Stanley 1979, Rosindell et al. 2010, Rosenblum et al. 2012), such that speciation is more analogous to colonization than to dispersal. Moreover, while phylogenetic data allow estimation of per capita rates of "diversification" (speciation – extinction) in a fairly straightforward way, parsing out speciation and extinction requires some hard-to-verify assumptions, and even fossil data are not entirely immune to this problem (Alroy 2008, Gillman et al. 2011, Rull 2013). In short, it is important to keep in mind that published speciation rates may often be underestimates and that some conclusions in the literature (described in this chapter) are likely to remain open to debate for some time given uncertain robustness of these conclusions to alternative assumptions underlying quantitative methods (e.g., Gillman et al. 2011, Rabosky 2012). I am not an expert on these methodological issues, so in the following sections I do not treat them in detail, but I do point readers to the relevant publications in which they have been raised.

10.3. EFFECTS OF SPECIATION ON COMMUNITY PATTERNS

In light of the difficulties of directly studying speciation in a community context, the remainder of this chapter is structured as follows. First, I address the hypothesis that speciation rate or the "time for speciation" is an important determinant of spatial patterns of species richness at various scales (see Sec. 10.4). This is the most direct link between speciation and its potential community consequences. In Section 10.5 I then discuss the species-pool hypothesis (see also Chaps. 3 and 6), which involves speciation only indirectly, as one determinant of regional species pools, the diversity of which determines local-scale diversity patterns.

10.4. HYPOTHESIS 7.1: SPATIAL VARIATION IN SPECIES DIVERSITY HAS BEEN GENERATED BY SPATIAL VARIATION IN SPECIATION RATES

As with dispersal, I have not stated a generic "speciation is important" hypothesis because, all else being equal, speciation *must* be an important determinant of large-scale diversity. However, speciation rate need not be an important predictor of patterns of species diversity if dispersal, selection, and drift (via immigration and extinction) are of overwhelming importance. For example, geographic isolation might favor an increased role for speciation in generating the regional species pool while simultaneously reducing diversity via limited dispersal, with the latter effect dominating the former (Desjardins-Proulx and Gravel 2012).

Here I address four predictions of Hypothesis 7.1. First, given the difficulty of matching the scales at which species diversity and speciation rates are quantified (speciation events can rarely be tied to a specific location), many studies address this hypothesis indirectly by asking whether speciation rates vary according to widespread predictors of species diversity (rather than species diversity itself), such as latitude (Predictions 7.1a and 7.1b). If speciation rates vary in concert with spatial patterns of species diversity, some causal link is likely. Other studies test for speciation rate as a potential explanation for exceptions to general patterns, or for marked differences in species diversity between ecologically similar but geographically distant habitats (Prediction 7.1c), or they focus specifically on the period of time during which speciation has had the opportunity to increase diversity under particular environmental conditions, rather than speciation *rate* (Prediction 7.1d).

Prediction 7.1a: Speciation rate increases as a function of island (or habitat island) area.

Methods 7.1a: Across a set of islands with at least some in situ speciation having occurred in the focal group of species, quantify species diversity, area, and speciation rate.

Results 7.1a: Many species-area relationships are quantified in settings where species diversity within a habitat patch or island has not been influenced by in situ speciation at all, such as plants in nearby forest fragments, or mammals in a small freshwater archipelago in previously glaciated regions (e.g., Lomolino 1982, Vellend 2004). In contrast, on isolated oceanic islands, in situ speciation can be an important source of diversity. For *Anolis* lizards on

Caribbean islands, Losos and Schluter (2000) found a species-area relationship driven largely by increased speciation rates on the largest islands. On islands of less than ~3000 km^2, molecular phylogenetic analysis indicated that none of the resident species arose via in situ speciation. On the four largest islands, species diversity was much higher, most species arose via in situ speciation, and the proportion of all species arising via speciation increased as a function of island area (Fig. 10.1a, b). Similar results have been found for *Bulimulus*

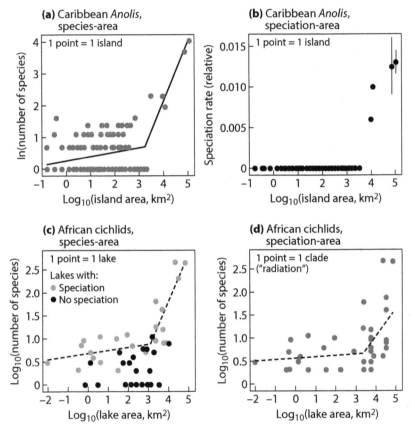

Figure 10.1. Relationships between species richness, speciation, and island or lake area. (a–b) Data from Losos and Schluter (2000) on lizards of the genus *Anolis*. In (b), error bars represent "the range of estimates resulting from different phylogenetic geographic reconstructions of the occurrence of ancestral taxa." (c–d) Data from Wagner et al. (2014) on cichlid fish in African lakes. There was no significant relationship between clade species richness and age, such that the relationship in (d) is driven largely by diversification rate. All lines come from two-slope regressions, shown only for lakes with speciation in (c). (No significant relationship was found for lakes with no speciation.)

snails on the Galápagos Islands (Parent and Crespi 2006, Losos and Parent 2009). For cichlid fish in African lakes, Wagner et al. (2014) found no influence of lake area on the *proportion* of species produced by in situ speciation, but a steep increase in the *number* of species produced by speciation with lake area, resulting in a steeper species-area relationship than expected based on species arriving only via dispersal (Fig. 10.1c, d).

Prediction 7.1b: Speciation rate is greater in the tropics than at higher latitudes.

Methods 7.1b: For a group of organisms known to show a latitudinal diversity gradient, estimate speciation rates at different latitudes.

Results 7.1b: The striking decrease in species diversity from the tropics toward the poles in many groups of organisms has fascinated biologists for centuries, and while phylogenetic studies generally support the prediction that *diversification* rate has been greater in the tropics (Mittelbach et al. 2007), the roles of speciation versus extinction are not always quantified. Studies finding decreased *speciation* rates with latitude include analyses of fossil data for planktonic foraminifera (Allen et al. 2006; Fig. 10.2a) and marine bivalves (Jablonski et al. 2006, Krug et al. 2009; Fig. 10.2b), as well as phylogenetic studies of amphibians (Pyron and Wiens 2013) and several mammal clades globally (Rolland et al. 2014). However, studies of New World mammals and birds find *lower* speciation rates in the tropics versus the temperate zone, with the latitudinal gradient thus explained by an even steeper latitudinal increase in extinction rate (Weir and Schluter 2007, Pyron 2014). Some researchers have called the latter results into question based on the possibility that faster molecular evolution in tropics leaves a false impression that sister species have been separated for longer in the tropics than they actually have (Gillman et al. 2011).

Prediction 7.1c: Seemingly anomalous patterns of species diversity—e.g., exceptions to the latitudinal gradient, or very different species diversity in similar but distant habitats—correlate with variation in speciation rates.

Methods 7.1c: In cases where a diversity anomaly has been identified, estimate speciation rates in the relevant regions.

Results 7.1c: Even in studies showing clear declines in species diversity with latitude for broadly defined clades, some subclades are the exception to the rule, showing diversity peaks elsewhere than in the tropics. For example, subclades of marine bivalves with extratropical diversity peaks have also been shown to have higher speciation rates at higher latitudes (Krug et al. 2007). In addition, many cases have been identified in which species diversity varies greatly between distant areas in otherwise similar habitats (Schluter and Ricklefs 1993), but such cases have only rarely been analyzed in a phylogenetic context. In the case of mangrove habitats, Ricklefs et al.

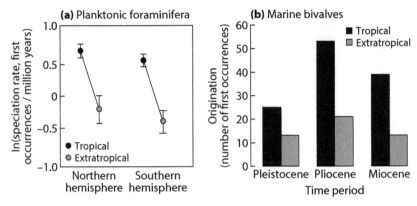

Figure 10.2. Rates of speciation or origination (of genera) in tropical and extratropical regions, as estimated using first occurrences of taxa in the fossil record for (a) foraminifera over the past 30 million years or (b) marine bivalves during different geologic epochs. Error bars in (a) indicate 95% confidence intervals. Data from (a) Allen et al. (2006) and (b) Jablonski et al. (2006).

(2006) found a correspondence between high plant species diversity and high speciation rates in the Eastern Hemisphere, and vice versa for the Western Hemisphere.

Prediction 7.1d: Spatial variation in species diversity is correlated with the time since the focal group of organisms first occupied different areas or habitats.

Methods 7.1d: Assess spatial variation in species diversity, and estimate the time at which species in the clade of interest first occupied different regions or habitats. The latter essentially involves reconstructing the habitat types of the common ancestors of extant species and using these as "character states" in phylogenetic analyses (Wiens et al. 2007).

Results 7.1d: This prediction has been pursued vigorously by John Wiens and colleagues (reviewed in Wiens 2011), with support coming from emydid turtles in different regions of North America (Stephens and Wiens 2003), from tree frogs between the tropics and the temperate zone (Wiens et al. 2011), and from plethodontid salamanders at different elevations in the tropics (Wiens et al. 2007; Fig. 10.3a) and in areas with different climatic conditions (Kozak and Wiens 2012; Fig. 10.3b). These results provide compelling evidence that time for speciation can be an important determinant of present-day patterns of species diversity (Hawkins et al. 2007), although doubt has been raised concerning the possibility that overly simplistic statistical methods obscure a possible effect of regional selection and drift in limiting diversity (Rabosky 2012).

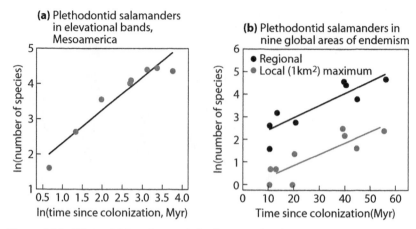

Figure 10.3. Effect of "time for speciation" on species richness of plethodontid salamanders (a) among 500-m elevational bands in Mesoamerica and (b) among global areas of endemism. Lines show best-fit linear least-squares regressions. Data from (a) Wiens et al. (2007) and (b) Kozak and Wiens (2012).

10.5. HYPOTHESIS 7.2: LOCAL SPECIES DIVERSITY IS ULTIMATELY DETERMINED BY THE PROCESSES THAT DETERMINE REGIONAL DIVERSITY (E.G., SPECIATION) RATHER THAN LOCAL-SCALE SELECTION AND DRIFT

Imagine an environmental gradient (productivity, climate, stress, etc.) along which species diversity varies. Why do more species occur under certain environmental conditions than others? In the decades after Hutchinson (1959) inspired ecologists to more thoroughly address the problem of species diversity, the majority of studies assumed that "more species exist in a given habitat because more species *can*, not because more species arose there" (Allmon et al. 1998). So, if local diversity is maximized at, say, intermediate productivity, perhaps constant selection is weakest and negative frequency-dependent selection strongest at intermediate productivity, such that the pattern would arise regardless of how many species in the regional pool were adapted to different levels of productivity. Alternatively, the predominance of intermediate productivity conditions over large spatial and temporal scales may have provided more space and time for species adapted to such conditions to accumulate. The latter is the "species-pool hypothesis" (Taylor et al. 1990, Zobel 1997), and while it involves speciation only indirectly, I include it here as one of the main ways in which speciation has been weaved into conceptual models and resulting empirical studies in community ecology.

The species-pool hypothesis is closely related to the "time for speciation" hypothesis (discussed earlier) and could be thought of as the "time and space for diversification" hypothesis. While conceptually fairly straightforward, this hypothesis is challenging to test, in part given the difficulty in defining "species pools" (Carstensen et al. 2013, Cornell and Harrison 2014). For example, there is no consensus on questions such as how to determine a most appropriate spatial scale for a species pool—whether to incorporate only spatial considerations (which species might have access to a given locality?) or whether to incorporate species-specific environmental tolerance (which species have access and can tolerate local abiotic conditions?). Many papers have weighed in on these issues (Zobel 1997, Srivastava 1999, Carstensen et al. 2013, Cornell and Harrison 2014), which I do not delve into here.

Prediction 7.2a: Local species diversity increases as a linear function of regional species diversity.

Methods 7.2a: For multiple regions or habitat types, estimate species richness at the local scale and the regional scale. If possible, also estimate factors potentially confounded with regional richness that might determine local richness.

Results 7.2a: There is a large body of literature on this prediction, most of which focuses not on whether the local-regional diversity relationship is just positive but whether the relationship is "saturating" (see Fig. 3.4b). Since by definition local species richness cannot exceed regional species richness, the two quantities are not statistically independent (e.g., if local sites all have zero species, regional richness must be zero). Researchers thus typically ignore the shape of the relationship near the origin (it is essentially always positive; Cornell and Harrison 2014) and focus on whether local richness levels off to a constant value as regional richness increases, that is, whether there is saturation. If local richness continues to increase across the full range of regional richness values, and especially if the relationship is linear, it is concluded that local species richness is dictated largely by the regional species pool (Cornell 1985), although this simple line of logic and associated statistical methods have been critiqued repeatedly (Srivastava 1999, Shurin and Srivastava 2005, Szava-Kovats et al. 2013, Cornell and Harrison 2014).

Of the two earliest tests of this prediction, one showed a pattern of saturation for birds on islands in the West Indies (Terborgh and Faaborg 1980; Fig. 10.4a) and the other showed no sign of saturation for gall-forming wasp communities on different species of oak (Cornell 1985; Fig. 10.4b). These were followed by dozens of case studies, with several reviews and meta-analyses concluding that a linear relationship (i.e., lack of saturation) was the most common pattern (Caley and Schluter 1997, Lawton 1999, Shurin and Srivastava 2005). A more recent study dealing more explicitly with the statistical nonindependence of the two scales of

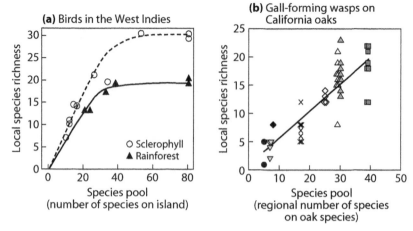

Figure 10.4. Relationships between local and regional (i.e., species pool) richness. (a) Breeding land birds on Greater and Lesser Antillean islands observed in local censuses in two types of habitat (y-axis) or across entire islands (x-axis); (b) cynipid gall-forming wasps in local populations of particular oak species (y-axis) or across the oak species range (x-axis). Each symbol type in (b) corresponds to a different oak species, with slight horizontal jittering of overlapping points. The lines in (a) approximate those shown in the original publication (fitting method unclear), and the line in (b) shows the best-fit linear least-squares regression. Data from (a) Terborgh and Faaborg (1980) and (b) Cornell (1985).

richness, however, concluded that patterns indicating saturation and lack of saturation were equally common, with many other tests inconclusive (Szava-Kovats et al. 2013). Either way, it seems clear that local species richness is very often (i.e., in at least half of the conclusive tests) linearly related to regional species richness, indicating fairly strong support for this prediction.

Some studies have analyzed not only regional richness as a predictor of local richness but also a series of environmental variables either potentially confounded with regional richness or themselves indicative of time and space for speciation. For example, Jetz and Fine (2012) analyzed data for vertebrates among global bioregions (i.e., biomes on different continents), finding that time-integrated bioregion area across the past 55 million years was a better predictor of species richness than just present-day bioregion area. Several studies with plants (Pärtel and Zobel 1999, Harrison et al. 2006, Grace et al. 2011, Laliberté et al. 2014) and birds (White and Hurlbert 2010) have included several environmental predictor variables and still found independent effects of regional richness on local richness. Overall, these results indicate that patterns of local-scale species richness can be strongly influenced by the processes that create the regional species pool, such as speciation.

Prediction 7.2b: The direction of local diversity-environment relationships can be predicted based on regional environmental history, with local diversity maximized under those conditions that have predominated over longer time periods in the region.

Methods 7.2b: In multiple regions with variable environmental histories, quantify species richness and environmental conditions in multiple localities.

Results 7.2b: The data requirements for testing this prediction are substantial, and I know of few relevant studies. One anecdotal example was the finding of a negative relationship between local bird species diversity and foliage height diversity in Patagonia (Ralph 1985), which ran counter to the finding of positive relationships in both eastern North America (see Fig. 8.5) and Australia (Recher 1969). One potential explanation was that the vegetation type with high foliage height diversity in Patagonia (*Nothofagus* forests) was regionally isolated and rare, thus providing less time and space for buildup of a diverse regional species pool (Ralph 1985, Schluter and Ricklefs 1993).

A more systematic test of this prediction was conducted by Pärtel (2002). Relationships between soil pH and local plant species diversity were compiled from across the globe, and the direction of these relationships was related to the soil pH (low or high) in the "evolutionary center" in the floristic region where the study was conducted. The evolutionary center was defined as the area of highest regional plant species richness. As predicted, positive local diversity-pH relationships predominated in regions with high-pH evolutionary centers, and vice versa (Table 10.1). In a similar analysis of local plant species diversity versus productivity, Pärtel et al. (2007) found a predominance of positive diversity-productivity relationships in the tropics, where highly productive growing conditions have been widespread for millions of years, and a predominance of negative relationships (or at least a decline from intermediate to high productivity) at higher latitudes, where highly productive conditions have been much rarer. Similarly, in an analysis of birds, mammals, and amphibians worldwide, Belmaker and Jetz (2012) found greater species richness at a "local" scale (~400 km^2) when local environmental conditions were more representative of the broader region (~30,000 km^2). In sum, while very few studies

TABLE 10.1. The Number of Studies Finding Positive or Negative Correlations between Local Plant Species Diversity and Soil pH in Floristic Regions with Their Evolutionary Center on Low- or High-pH Soils

		Direction of Diversity-pH Correlation	
		Positive	*Negative*
Evolutionary center	**High pH**	39	9
	Low pH	13	21

Source: Data from Pärtel (2002).

of this nature have been conducted, they provide compelling evidence that the explanation for local diversity-environment correlations resides at least partly in processes such as speciation that have created different numbers of species adapted to different environmental conditions.

What are the mechanistic underpinnings of speciation? The causes of speciation are the focus of a massive body of literature, summarized in several books (Schluter 2000, Coyne and Orr 2004, Nosil 2012). Many empirical studies explore the influence of environmental conditions or key organismal traits on rates of speciation (McKinney and Drake 1998, Magurran et al. 1999, Coyne and Orr 2004). As one example of direct relevance to Prediction 7.1b, higher temperatures and thus greater energy availability have been hypothesized to result in faster molecular evolution and therefore increased speciation rates and species diversity (Rohde 1992; Allen et al. 2002, 2006; Wright et al. 2003; Davies et al. 2004; Mittelbach et al. 2007). Some empirical studies support this hypothesis (Davies et al. 2004, Gillman and Wright 2014), although results are also consistent with the possibility of causality in the reverse direction: speciation may be accompanied by genetic bottlenecks or unusually strong selection and thus bursts of rapid molecular evolution (Dowle et al. 2013).

Is evolutionary diversification (speciation minus extinction) diversity dependent and therefore self-regulating? This question is the large-scale analogue of Prediction 6.1a, asking whether local diversity is limited by the rate of input via dispersal or, instead, by selection and drift that keep local diversity more or less constant (see Box 10.1). Several lines of evidence, such as decelerating rates of diversification over time or independent variation between clade age and richness, suggest that high regional species diversity feeds back to slow diversification (Rosenzweig 1995, Rabosky 2013). Other empirical studies (e.g., see Prediction 7.1d) identify cases in which diversity-dependent feedback, if present, has not yet had a dominating influence on clade richness (Wiens 2011).

10.6. SUMMARY OF EMPIRICAL STUDIES OF SPECIATION IN ECOLOGICAL COMMUNITIES

As with most hypotheses and predictions presented in this book, sometimes empirical evidence is supportive, and sometimes it is not (Table 10.2). Relative to the predictions stated in previous chapters, few studies have tested the predictions in this chapter (possibly with the exception of local-regional richness relationships). Nonetheless, the studies that have been conducted clearly indicate that the rate or time for speciation can limit species richness either at a regional scale or in particular habitats (e.g., elevational zones). Considerable

TABLE 10.2. A Summary of the Empirical Support for, and Challenges and Caveats Involved with, Hypotheses and Predictions Based on the Importance of Speciation and Species-Pool Effects in Ecological Communities

Hypothesis (H) or Prediction (P)		Empirical Support	Challenges and Caveats
H7.1	**Speciation rate determines species diversity**	Speciation rate and time for speciation clearly contribute to limiting large-scale species richness in some cases; selection and/or drift dominate in other cases	Speciation is especially difficult to estimate, so results and conclusions are frequently revisited in light of alternative model assumptions
P7.1a	Speciation-area relationship	Clear evidence from a very few cases studies	Relevant in a narrow range of situations: many discrete areas unlikely to have any species that arose via in situ speciation
P7.1b	Speciation-latitude relationship	Higher speciation appears to play a role in driving high tropical diversity in some cases; some counter-examples	Persistent uncertainty in assumptions underlying models used to estimate speciation rates
P7.1c	Speciation rates correlate with anomalous diversity pattern	Some compelling examples	Persistent uncertainty in assumptions underlying models used to estimate speciation rates
P7.1d	Time for speciation effect	Some compelling examples	Robustness of results to alternative modeling frameworks has been questioned
H7.2	**Species pool size determines local species diversity**	Compelling evidence that local diversity is often limited by regional species pool diversity, rather than by selection and/or drift	
P7.2a	Linear local-regional richness relationship	At least half of conclusive tests are consistent with this prediction	Methods frequently critiqued; many studies do not incorporate potentially confounding variables
P7.2b	Environmental history predicts direction of diversity-environment relationship	Very few tests, but convincing evidence in those cases	High data requirements; difficult to establish a basis for predicting the direction of relationships

debate surrounds the question of how often selection and drift (i.e., "ecological limits") render the rate of speciation irrelevant to predicting diversity (Wiens 2011, Rabosky 2013), and the next decade or so promises far more data and methodological development with which to resolve this debate.

Patterns of local diversity are also often quite clearly influenced by the sizes of species pools adapted to different conditions. Evidence comes from studies in single regions where habitat-specific species pools can predict spatial variation in species richness (see Prediction 7.2a), and more convincingly from comparative studies across regions in which a priori predictions can be made concerning the direction of diversity-environment relationships (Prediction 7.2b), although the very small number of such studies makes it difficult to generalize. Speciation clearly cannot be ignored in community ecology.

PART IV

CONCLUSIONS, REFLECTIONS, AND FUTURE DIRECTIONS

CHAPTER 11

From Process to Pattern and Back Again

11.1. THE RELATIVE IMPORTANCE OF
DIFFERENT HIGH-LEVEL PROCESSES

In the preceding chapters, I articulated a theory of ecological communities based on the expected consequences of four high-level processes and subsequently evaluated the empirical evidence that these processes operate in nature. The main overarching conclusion from the empirical evidence (Chaps. 8–10) is that *all high-level processes can be important determinants of community structure and dynamics*. This simple and seemingly obvious conclusion has several important consequences. First, in the absence of a priori knowledge of a particular community, one cannot exclude the possibility that selection (in any of its forms), drift, dispersal, or speciation has an important influence on its structure or dynamics. Second, when community ecologists say that they are "testing theory" or "testing a hypothesis," their domain of application is almost always system specific. It is exceedingly rare that we universally reject a theory or hypothesis. Support for, or rejection of, a hypothesis applies only to the system under investigation, and there is no way to know whether the conclusions apply more broadly except by studying other communities. This fact makes the falsifiability criterion for science (Popper 1959) of limited value in ecology (Pickett et al. 2007). Finally, as is the case across the biological sciences, community ecology is often concerned with the question of the *relative* importance of different processes (Beatty 1997). The importance of selection versus drift, for example, in setting the trajectory and outcome of local community dynamics depends on whether the objects of study are plants in a grassland, damselflies in a lake, beetles in a lab container, or birds in a forest.

We can pose the question of relative importance at multiple levels. Many studies described in Chapters 8–10 posed the question within individual systems. Strong negative frequency-dependent selection dominates the dynamics of some annual plant communities, at least over the short term (Levine and HilleRisLambers 2009), while selection among co-occurring damselflies in freshwater appears to be very weak, allowing for an important role of ecological drift (Siepielski et al. 2010). Importantly, in neither case can we conclude that drift or selection, respectively, is entirely absent. We can also assess relative importance by considering empirical studies across a wide range of systems, evaluating how frequently different processes are found to be locally important. At present, we can do this only in a qualitative, and decidedly subjective, way (see summary tables in Chaps. 8–10). With those qualifiers in mind, my overarching interpretations of the evidence are as follows:

- Spatially and temporally variable selection has a major influence on most if not all ecological communities, contributing to the maintenance of species diversity and generating patterns of beta diversity.
- Local negative frequency-dependent selection (i.e., not an emergent property of spatially variable selection) is likely occurring among many pairs of species in many places. The logistical challenges involved in studying this process directly, and the consequent paucity of studies doing so, makes it difficult to say anything more specific.
- Spatially and temporally variable selection, as well as negative frequency-dependent selection, appears to be of widespread importance, but in many systems (i.e., too many to classify as unusual exceptions) environmental change can also trigger positive feedbacks. Such feedbacks cause unusually rapid community responses via positive frequency-dependent selection and resistance to change in the reverse direction when the trigger itself is reversed.
- Given the strength of various forms of selection, ecological drift is unlikely to be a process of major importance in most places or times. However, drift can certainly have an important influence on the dynamics of some sets of species under particular conditions, especially at small spatial scales. The possibility of very weak selection (and therefore an important role for drift) among many of the species in hyperdiverse communities such as tropical rainforests and coral reefs remains a viable theoretical possibility.
- For many if not most species, spatially limited dispersal prevents occupancy of localities where they could otherwise persist. Thus, local species diversity is often limited by the rate of incoming dispersal (immigration), and beta diversity is influenced by dispersal. To the extent that the presence

of particular species alters local selection dynamics among other species, dispersal can lead to any number of downstream effects on community composition (details will be highly case specific and can be understood with respect to alteration of other high-level processes).

- Speciation is obviously a key process underlying the generation of regional species pools, and in some cases the rate of speciation limits regional diversity. In other cases, regional pools appear saturated via selection and drift, such that there is no speciation-diversity relationship, but to date too few studies have been conducted to assess the frequency of these different outcomes.

- Local-scale variation in species diversity along various environmental gradients (e.g., stress, productivity) may very often be a result of species pools of different diversities having evolved under different conditions, rather than short-term consequences of local selection and drift. The relative importance of these different possibilities across systems and at different scales cannot be assessed at present.

11.2. PROCESS-FIRST AND PATTERN-FIRST APPROACHES TO COMMUNITY ECOLOGY

The approach taken in this book has been decidedly "process-first": if process X is an important determinant of community structure and dynamics, what do we expect to see in natural or experimental communities? However, as discussed in Chapters 2–4, this is not the only approach to community ecology, and probably not even the most common. For centuries, naturalists have been noting intriguing patterns in ecological communities (Lomolino et al. 2010), the quantification of which represents the start of the "pattern-first" approach: how strong/common is pattern Y, and what processes might have caused it? Patterns receiving especially large amounts of attention include species-area relationships, relative abundance distributions, diversity-productivity relationships, the distance decay of compositional dissimilarity, and latitudinal diversity gradients (Ricklefs and Miller 1999, Krebs 2009, Lomolino et al. 2010, Mittelbach 2012).

I want to make two points clear concerning these different approaches:

1. Neither the process-first nor the pattern-first approach to community ecology can be deemed objectively "better." To the degree that they share the same goal—understanding how processes produce patterns—both are of critical importance.

2. The four-process framework of the theory of ecological communities is equally applicable and useful in both cases.

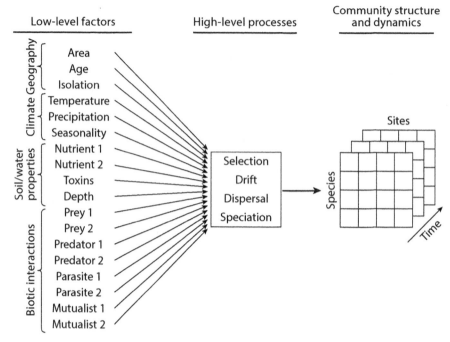

Figure 11.1. The theory of ecological communities based on high-level processes sits at the nexus of the process-first approach (starting from low-level factors) and the pattern-first approach to community ecology. This figure has been modified from Figure 4.3 by adding geographic variables to the low-level factors, listing the four high-level processes (not only forms of selection), and including a temporal component to community patterns.

How is the framework presented in this book relevant to the pattern-first approach to community ecology? Essentially, high-level processes provide the key link between low-level factors and the patterns we observe in nature, so we must work through them regardless of which side we start from (Fig. 11.1). I hope to have demonstrated so far that the seemingly innumerable and indigestible array of theoretical ideas and models found in any given community ecology textbook can be rendered digestible and conceptually coherent by relating each one to a small set of high-level processes, each of which has universally applicable consequences (Fig. 11.1; see also Fig. 4.3). Importantly, the ideas and models of which I speak include not only process-first models exploring the consequences of species interactions, stochastic fluctuations, and dispersal (these have dominated the book) but also post hoc candidate explanations for widespread patterns in nature, of which there are a great many. By some counts, there are at least 120 such explanations for patterns of species diversity (Palmer 1994), and 32 just for the latitudinal diversity gradient (Brown 2014).

Most post hoc explanations for ecological patterns have been expressed verbally rather than mathematically, often with rather vague logical links. Using the latitudinal gradient as an example, commonly invoked causal factors include historical disturbance (e.g., glaciation), productivity, environmental stress, climatic stability, environmental heterogeneity, and species interactions (Lomolino et al. 2010). On their own, only one of these is even close to being self-explanatory: higher environmental heterogeneity ought to support greater species diversity via spatially variable selection. Otherwise, it is not clear how these factors might relate to species diversity per se. To quote from a widely used ecology textbook (Krebs 2009), "Climate determines energy availability, and the key variables for terrestrial plants and animals are solar radiation, temperature, and water. Climates that are more stable are more favorable, cause higher productivity, and all these factors work together to support more species." This "explanation" is difficult to make sense of. It says only that many individual plants or animals might thrive best when conditions are consistently warm and moist, but it says nothing about why more *species* should be found under such conditions. I think that articulating how such low-level factors might be expected to influence the high-level processes that generate diversity (speciation, dispersal) or those that help maintain or reduce diversity (selection and drift) would be of great benefit to the conceptual coherence of community ecology. This approach, in turn, could facilitate student understanding of the entire pathway from low-level processes to patterns in nature. To be sure, many authors have done just this (e.g., Ricklefs and Schluter 1993b), but it is not routine, and I think that the theory of ecological communities can be of considerable value as the nexus of different approaches (Fig. 11.1).

To summarize, the theory of ecological communities does not advocate any particular methodological approach to community ecology. Rather, it has led me to argue for the collective benefits in understanding that can be gained by formally articulating hypotheses connecting low-level processes and patterns (regardless of where one begins) using the conceptual framework of four high-level processes.

11.3. THE CURIOUS CASE OF MACROECOLOGY

Building on conceptual developments in integrating local and regional processes (Ricklefs 1987, Ricklefs and Schluter 1993a, Leibold et al. 2004, Holyoak et al. 2005), this book explicitly considers that community patterns observed over large scales of space (e.g., continents) or time (e.g., millions of years) fall under the same conceptual umbrella as the small-scale studies that have traditionally defined community ecology. Ecological studies at the largest spatial scales are often described as "macroecology," which includes both comparative studies across species (e.g., body size and geographic range) as well as some community-level patterns (e.g., species diversity) (Brown 1995, Gaston and Blackburn 2000, McGill 2003a). In one sense, macroecology (specifically the portion that covers community-level patterns) falls comfortably within the pattern-first approach described previously and can therefore be related to the theory of ecological communities in a straightforward way. Studies of latitudinal diversity gradients (described earlier) represent a case in point.

However, some theoretical approaches in macroecology take an entirely different approach, which I am unable to relate to the theory of ecological communities. This situation is of special interest because such approaches concern horizontal communities and thus might be expected to fit comfortably in the framework presented here. An illustrative example is the "maximum entropy theory of ecology" used to predict species abundance distributions (Harte 2011, Harte and Newman 2014). The core of this theory works as follows: in a given locality with area A, containing S species, J individuals, and with total metabolic rate M, species relative abundances should follow a distribution that requires no additional information to predict. The "no additional information" part is the origin of the name "maximum entropy." Information theory states that the amount of information in a set of objects is minimized when a quantity called *entropy* is maximized (entropy is calculated using the same formula as the well-known Shannon diversity index $= \Sigma_i p_i \times \log(p_i)$, with p_i as the relative abundance of species i) (McGill and Nekola 2010). So, the relative abundance distribution is predicted as the set of p_i's that maximize entropy, subject to the constraints A, S, J, and M. That might sound mysterious, but I know of no more simple or intuitive way to explain it.

It is certainly intriguing that predictions of maximum entropy theory are well matched by data (Harte 2011), and it strikes me as worth pursuing the question of why this is the case. Indeed, this approach shares with the theory of ecological communities the aim of taking a complex set of low-level processes or phenomena and condensing them into something simpler. However, because the model is described as working "without invoking any specific ecological mechanisms" (Harte et al. 2008), I have no idea how to make sense of the successful predictions with respect to other ecological theories, including my own. These results can be seen as being aimed purely at prediction, regardless of mechanism, which is a valid and important scientific goal (Peters 1991); however, in most natural systems I can think of, it would require as much or more effort to accurately estimate A, S, J, and M as it would to estimate all the p_i's, so I'm not sure how useful these are as predictions. And since the theory is explicitly mechanism free, it is also unclear to me whether or how these results deepen our understanding of fundamental ecological questions. Other "mechanisms" invoked in macroecological studies (reviewed in McGill and Nekola

2010) that fall outside the scope of the theory of ecological communities include the central limit theorem and fractal geometry.

11.4. THE THEORY OF ECOLOGICAL COMMUNITIES
COVERS (ALMOST) EVERYTHING

My main reason for explicitly mentioning these macroecological ideas is not to
imply that they are any less useful or less important than other ideas or models
(although, clearly, I have doubts in some cases). Rather, I wanted to acknowledge the one major exception I can think of in terms of conceptual constructs
for horizontal communities that can be accommodated by the theory of ecological communities. The vast majority of theories, ideas, models, and hypotheses
in community ecology consider some combination of what I call low-level and
high-level processes and thus fit comfortably under the broad theory presented
here, regardless of whether the initial emphasis was on process or on pattern
(see Table 5.1). But there are at least a few exceptions.

CHAPTER 12

The Future of Community Ecology

At several junctures during this book I have argued that one of the main contributions of the Theory of Ecological Communities is to facilitate a clearer and more precise understanding of process-pattern linkages. Selection, drift, dispersal, and speciation have broadly applicable consequences in ecological communities, and all but a very few explanations for community dynamics and patterns can be understood with reference to one or more of these interacting high-level processes.

This theory may be viewed by some as self-evident. However, in the twenty-first century, the same is true of the theory of evolution by natural selection (at least among scientists), which nonetheless remains rather useful in evolutionary biology (see also Chap. 5). In both cases, the contemporary value of the theory resides to a large extent in its providing a general, fairly simple, and conceptually coherent framework that allows us to readily identify commonalities between populations or communities that might otherwise differ profoundly in their details. To reiterate an argument from Chapter 4, whether fitness depends on body coloration that permits camouflage from predators (Kettlewell 1961) or on having a beak size that matches seed availability (Grant and Grant 2002), adaptive evolution occurs via natural selection. Likewise, whether rare species gain an advantage owing to an abundance of unused resources (Tilman 1982) or to a paucity of natural enemies (Connell 1970, Janzen 1970), species coexistence can be maintained via negative frequency-dependent selection (Chesson 2000b, HilleRisLambers et al. 2012).

While the theory of ecological communities can facilitate understanding, it is not self-evident that it leads to specific new research directions. At this point in many books or papers of this nature, authors direct the next generation of graduate students to the research projects they should now do (with most of them being projects about to be undertaken by the authors themselves). As a committed pluralist regarding different approaches to ecology, I hesitate to get prescriptive. Thus, the following list of future research directions can be read as a highly personal perspective on things. In the process of writing this book, I drew upon many hundreds of papers and books. Along the way, and in light of a focus on high-level processes, many ideas came to mind. Most of these ideas have already been expressed in some form in the literature, so to a large extent I am simply pointing out nascent research directions that seem especially promising to me (most of which I do not intend to pursue myself).

12.1. SOME MISSING META-ANALYSES

In attempting to review the literature on many different individual topics, I have found papers reporting systematic reviews or meta-analyses immensely useful. We know very well that the answer to most ecological questions varies from place to place and at different scales, so advancing our general understanding depends on systematic evaluations of the frequency, strength, and context dependence of certain relationships. For example, in place of reading dozens of individual studies reporting relationships between local and regional species richness, I could begin with just a few reviews and meta-analyses to get a sense of how often different results are found (Shurin and Srivastava 2005, Szava-Kovats et al. 2013, Cornell and Harrison 2014) and then consult selected case studies (e.g., Terborgh and Faaborg 1980, Cornell 1985) to get a sense of the raw data.

For some topics, I was surprised to find no such systematic reviews or meta-analyses (although it is certainly possible I missed some important papers). As far as I can tell, the following topics have either never been subject to a systematic meta-analysis, or at least not in the past 10–15 years, during which the number of primary studies on almost all topics in community ecology has increased greatly:

- The relative strength of density- or frequency-dependent interactions within versus between species (Prediction 2a). The reviews and meta-analyses of which I am aware (Connell 1983, Schoener 1983a, Goldberg and Barton 1992, Gurevitch et al. 1992) are at least 20 years old, with a great many empirical studies and methodological advances having occurred since then (see Chap. 8).

- The influence of patch area on beta diversity (Prediction 5a-(ii)).
- The influence of community size on the strength of composition-environment relationships (Prediction 5a-(iii)).
- The relationship between species richness and habitat patch (or island) isolation (Prediction 6.1a).

12.2. COORDINATED, DISTRIBUTED EXPERIMENTS (OR OBSERVATIONAL STUDIES)

Meta-analyses are far from perfect. As every field ecologist knows, each empirical study involves a great many decisions about design. For example, in conducting plant community surveys, we decide on the number of study plots, their size and shape, their distribution in space, the time of year during which we make observations, the number of times we survey each plot, the scale used to quantify abundance, the species we will include (e.g., all plants, only vascular plants, only trees), the amount of lumping of difficult-to-identify taxa, which environmental variables to measure, and how to measure them. Some of these individual decisions might be of a fairly minor nature, but collectively they can add up to make two seemingly similar studies quite difficult to compare quantitatively. This challenge can stifle efforts to assess the generality (or lack thereof) of patterns and processes across different regions. Meta-analyses deal with this difficulty by incorporating potentially confounding sources of study-to-study variation, but most often, the comparability of different studies remains somewhat uncertain.

A solution to this challenge is to implement a standardized methodology in many different places at once. This approach that has been dubbed "coordinated distributed experiments" (Fraser et al. 2012, Lessard et al. 2012), although the approach is equally applicable to observational studies. Few ecologists have the budget for such a project, but several recent studies have taken a creative approach to getting this done by enlisting many different researchers to implement a design of sufficient simplicity that individual researchers need little if any additional funding to join in.

An especially impressive example is the Nutrient Network (NutNet), a network of more than 75 grassland sites across the globe, initially designed to test the effects of nutrients and herbivory on community and ecosystem properties via manipulative experimentation (Borer, Harpole, et al. 2014; Borer, Seabloom, et al. 2014). With yearly observations of community dynamics in unmanipulated communities as well, a wide variety of other questions have been addressed, ranging from the relationship between species diversity and productivity (Adler et al. 2011) to the causes of exotic species invasions (MacDougall et al. 2014). Almost any local-scale question in community ecology could be fruitfully approached using controlled distributed experimentation or observation. In addition, testing for signatures of the action of high-level processes (reviewed in Chaps. 8–10), rather than low-level processes, might provide an opportunity for studies of this type to generalize beyond individual habitat types.

12.3. EXPERIMENTAL TESTS OF THE CONSEQUENCES OF (EFFECTIVE) COMMUNITY SIZE

In systems where recognizing discrete habitat patches is fairly straightforward (e.g., forest fragments, ponds, rock pools), such patches typically vary greatly in area or volume and therefore in community size (J). Human land-use activities are also a frequent cause of reduction in habitat patch area. Thus, understanding the consequences of altered community size is of widespread ecological importance. However, the size of a habitat fragment is correlated with many factors other than community size, most notably edge effects and environmental conditions (Harrison and Bruna 1999, Laurance et al. 2002, Fahrig 2003). In other words, small habitat patches might be more prone not only to drift (given reduced J) but also to different forms and strengths of selection (see Chap. 9). A similar argument applies to temporal environmental fluctuations, which simultaneously reduce effective community size and cause temporally variable selection (see Chap. 8).

Isolating the effect of community size can be approached by controlling for environmental correlates of habitat area statistically (e.g., Ricklefs and Lovette 1999), but whether all important variables have been measured will always be open to question. In principle, experimental manipulation ought to be able to disentangle the different causal factors, but even experimentally created small patches involve strong edge and environmental effects, as well (Debinski and Holt 2000). The time seems ripe for clever experimenters to solve this problem.

For a zooplankton community, one could imagine using mesh to enclose a certain volume of water in a lake—say, 1 m × 1 m × 25 cm—with 16 internal compartments, each 25 cm on a side. If each compartment was fully sealed, these would be closed communities of "small J." Medium-J communities—50 × 50 × 25 cm in volume—could be created by poking holes in the internal walls of compartments of that area, and likewise for large-J communities (the entire 1 m × 1 m × 25 cm). The hope (to be verified by a limnologist) is that the holes allow movement among internal compartments without altering edge effects introduced by the mesh itself, which otherwise stays in place. This design obviously bears a strong resemblance to a dispersal-manipulation experiment, but with an explicit eye toward isolating the influence of community size. Better experimental ecologists than I can no doubt come up with some other new and creative designs.

12.4. EXPERIMENTAL REDUCTION OF IMMIGRATION IN THE FIELD

Just as community size is often studied via the effects of habitat patch size, the community consequences of dispersal are often studied via the effects of patch isolation. And as in the case of area, isolation might be confounded with other patch characteristics (see Chap. 9). Experiments can manipulate dispersal directly or indirectly, but with some limitations. In artificial lab systems, we can allow for a wide range of dispersal levels among habitat patches (e.g., via tubes connecting small bottles). In the field, however, all we can do, for the most part, is augment immigration. For plants, seed-addition experiments show that local species diversity is often limited by immigration, but what would happen if we reduced immigration? Brown and Gibson (1983) described just such a thought experiment—the construction of a dispersal-proof fence around a given ecological community—to test the degree to which local community properties depend on dispersal (see also Holt 1993).

Both reducing and augmenting immigration to varying degrees in the field would allow a quantitative assessment not only of the relevance of dispersal but of the shape of relationships between community properties (e.g., species diversity) and immigration. I know of few if any such experiments, likely due to the logistic difficulties involved (note that I always hedge with "few if any," given no claim to comprehensive knowledge of the literature). For a grassland plant community, it ought to be relatively straightforward to build mesh enclosures that keep out the majority of incoming seeds. By first assessing seed shadows (proportion of seeds moving different distances) of different species, one could then simulate both the natural rate of seed input per year, as well as reductions (all the way to zero) and increases. Enclosures no doubt have an influence of their own, but at the very least they can be standardized across treatments, and for some portion of the growing season (when no seed is being produced) they could be removed. Again, I expect that others could come up with more clever designs than this one, but the core question would remain the same: what are the consequences of reduced immigration?

12.5. SYNTHESIZING STUDIES OF SPECIES COEXISTENCE AND SPECIES DIVERSITY

At first glance, the topics of species coexistence and species diversity are so closely connected as to appear nearly identical (Huston 1994, Tokeshi 1999). Indeed if co-occurrence is synonymous with coexistence, then this will be the case. If one site has more species than another, then more species co-occur and therefore coexist in that site. However, according to contemporary use of the term, species coexistence implies that each species in the community will tend to rebound from low relative abundance (Chesson 2000a). This more strict definition of the concept of coexistence clearly decouples it from the concept of species diversity. One site might have more species than another not because of stronger negative frequency-dependent selection (i.e., coexistence mechanisms) but because it has more species present as demographic sinks (i.e., immigration) or because the species pool contains a larger number of species that can tolerate the local abiotic conditions (see Chaps. 9–10).

Although I think that many, if not most, researchers in community ecology appreciate this distinction, it can still lead to debaters talking past one another. For example, as described also in Chapter 5, Fox (2013) has deemed the intermediate disturbance hypothesis unsound on mathematical grounds, because while disturbance might slow competitive exclusion, on its own it doesn't create the negative frequency-dependent selection needed for stable coexistence. This is a valid argument. However, Huston (2014) countered by making the equally valid argument (among others) that predicting spatial and temporal variation in species diversity is not the same thing as predicting long-term stable species coexistence. Empirical studies increasingly acknowledge this distinction (e.g., Laliberté et al. 2014), but I think a "modern synthesis" of sorts, bringing together studies of species coexistence and studies of species diversity, is overdue and can help reconcile different points of view such as those of Fox (2013) and Huston (2014). Perhaps this book can provide one step in that direction. The fact that coexistence in species-rich communities appears to be inherently "high dimensional," involving many different traits and trade-offs simultaneously (Clark et al. 2010; Kraft, Godoy, et al. 2015), suggests that focusing on selection as a high-level process, rather than on particular low-level mechanisms, can facilitate conceptual synthesis.

12.6. COMMUNITIES AND ECOSYSTEMS AS COMPLEX ADAPTIVE SYSTEMS: LINKING COMMUNITY PROPERTIES TO ECOSYSTEM FUNCTION

This book is based, in large part, on the analogy between the dynamics of alleles in populations (as described by models in population genetics) and of species in communities (as described by models in community ecology). In both cases, biological variants are described based on a set of discrete categories: an allele can be *A* or *a*, and a species can be *Acer saccharum* (sugar maple) or *Fagus grandifolia* (American beech). Quantitative genetics provides an alternative way to model evolutionary change within species by focusing on heritable phenotypes (e.g., beak depth), which can take on a continuous range of values (see also Chap. 5). To the extent that we can measure phenotypes in a comparable way across species, models inspired by quantitative genetics can be used to describe the dynamics of community-level trait distributions (Norberg et al. 2001, Shipley 2010). Some simple examples of this approach were discussed in Chapters 5 and 8 (see Predictions 1b and 2c).

Viewing community change as a quantitative genetics problem flows out of the broader view of communities and ecosystems as "complex adaptive systems" (Levin 1998). In a complex adaptive system, "patterns at higher levels emerge from localized interactions and selection processes acting at lower levels" (Levin 1998). From this perspective, we can think of the total rate of biomass production (the most commonly measured "ecosystem function") as emerging from localized interactions between individuals of different species and selection among them (Norberg et al. 2001, Norberg 2004). In addition, just as we expect genetic variance to allow populations to "track" environmental change (Fisher 1958), so might we expect species diversity to allow productivity to be maximized as the environment changes (Norberg et al. 2001).

In Chapter 8, I described studies showing that productivity generally increases with increasing species diversity in a local community as evidence of negative frequency-dependent selection (see Prediction 2a). These experiments have been used to argue that global biodiversity loss has compromised the functioning of ecosystems (Cardinale et al. 2012) and will continue to do so, although this conclusion has been called into question given that *at the local scale*, barring wholesale habitat conversion (e.g., rainforest to corn field), species diversity has not tended to show directional changes over time, and when it has, increases have been as likely as decreases (Vellend et al. 2013, Dornelas et al. 2014, Elahi et al. 2015, McGill et al. 2015). It is clear, however, that many local communities have experienced tremendous turnover in species composition over time (Dornelas et al. 2014, McGill et al. 2015). To the extent that such changes have been "adaptive," we might expect that they have served to maintain ecosystem function as the environment has changed.

This prediction is testable via experimentation. A given community (e.g., plants in a plot of grassland) can be subject to an environmental change (i.e., external selection), or not, with community composition allowed to reach a new quasi-equilibrium state. The resulting community states (in altered vs. control environments) can then become levels of a treatment in a second experiment. That is, in the second experiment initial abundances are set according to one of the two quasi-equilibrium states reached in the first experiment, and these initial community-state treatments are crossed with the two types of environment (altered vs. control). The prediction based on the view of communities as complex adaptive systems is that productivity will be maximized when community composition "matches" the environment. Any number of questions could be added, including the strength of selection (magnitude and rate of environmental change), trait variance or species diversity in the initial community, and the dimensionality of selection (i.e., single vs. multiple variables changing simultaneously).

12.7. QUANTIFYING RELATIVE IMPORTANCE

I began Chapter 11 with an assessment of the relative importance of different high-level processes across different types of natural communities, but the results were rather unsatisfying in that they were entirely qualitative and replete with uncertainty. Within a particular local community, several methods can be used to assess the relative importance of different processes or factors, but with important limitations in each case. For example, we can experimentally manipulate a few factors of interest to test the relative importance of, for example, dispersal (seed addition) and particular agents of selection (nutrient addition) in determining local species richness. But the magnitude of the statistical effect will depend very strongly on how far we "push" each factor, and there's no obvious way to convert each effect to common units: Δ richness/(g seed) or Δ richness/(species added) versus Δ richness/(g nutrient). Similar problems apply to observational studies using multiple regression-type analyses to assess the relative importance of different predictors of species diversity or composition, with the added problem that the interpretation of some variables (e.g., spatial proximity of sites) is far from clear (see Chaps. 8 and 9).

Some researchers have developed common currencies for expressing the relative importance of different processes. Studies of species coexistence, for example, first fit models to data on species responses to various factors (e.g., intra- and interspecific densities) and then estimate constant selection ("fitness differences") and negative frequency-dependent selection ("niche differences") in units of low-density relative growth rates (Chesson 2000b, Adler et al. 2010). These are important efforts, but they are quite narrowly focused on the question of stable species coexistence, which is just one of many outcomes or patterns of interest in community ecology.

Comparing the relative importance of particular processes across systems seems more straightforward, at least in principle. For example, coordinated distributed experiments (see Sec. 12.2) could be first applied to quantifying, in many different places, the degree to which local species richness is limited by immigration from the regional species pool. If relative importance across sites was also assessed for other high-level processes—for example, constant and negative frequency-dependent selection, as described in the previous paragraph—one could then use the comparisons *across* sites to assess relative importance *within* sites. For example, the effect of immigration in a particular site may be high (relative to other sites) while the strength of negative frequency-dependent selection is low (relative to other sites). This approach would then essentially be assessing the relative importance (within sites) of relative importance (among sites).

In short, community ecologists are very often interested in assessing the relative importance of different processes in determining community properties both within and among different sites, but the methods for doing so all have

important limitations. New approaches to this problem would be most wel-
come, although this is unlikely to be an easy problem to solve—that is, "short-
cuts" rarely seem to work in ecology (Fox 2012).

12.8. DEVELOPING A CORE SET OF COMMUNITY MODELS BASED ON HIGH-LEVEL PROCESSES

Community ecology is already rich in mathematical models, but it is not always straightforward to relate one model to the next. Population-genetic models have the appeal of more or less always being based on four key parameters:
N (population size), s (selection coefficient), m (gene flow), and μ (mutation)
(Hartl and Clark 1997), which makes it relatively easy to compare models with
one another. Ecological models of communities could likewise focus on four
key parameters: J (community size), s (selection coefficient), m (dispersal), and
v (speciation rate). Ecological neutral theory (Hubbell 2001) does just this, but
without selection. Chapter 6 presented simulation models including all four
processes, and analytical versions of these models are possible in many cases.
Some authors have begun to construct general analytical community models
that effectively merge selection and neutral theory (Haegeman and Loreau
2011, Noble and Fagan 2014). Continuing this effort, with the aim of eventually distilling the set of analytical models in horizontal community ecology to
a manageable set based on high-level processes, would be of great service to
theory-oriented students of the field.

One key difference between the modeling traditions in population genetics
and ecology is the assumption of constant N in population genetics versus the
treatment of J as an outcome of density-dependent processes among multiple
species in ecology (Vellend 2010). However, since the assumption of constant
N seems to have developed in population genetics largely for the sake of mathematical convenience, rather than as a representation of natural populations
(Lewontin 2004), development of ecological models based on high-level processes is likely to have crossover benefits for theory in population genetics, as
well (e.g., Ellner and Hairston Jr 1994).

12.8. A SYNTHESIS OF SYNTHESES IN COMMUNITY ECOLOGY?

In Chapter 2, I circumscribed the domain of the theory of ecological communities to include largely what we can think of as "horizontal" sets of species
(see Fig. 2.1). Species that share similar constraints on fitness (e.g., in terms of
resource needs and the types of enemies or mutualists with which they interact)
are theoretically analogous to genetic variants within species. It was therefore

possible to define a core set of high-level processes for horizontal communities that mirrors the core set of processes in population genetics, even if some of the details differ considerably (e.g., mutation vs. speciation). This raises the question of whether an even broader synthesis is possible, combining horizontal communities, food webs, and mutualist networks into a single framework. I don't have an answer to that question, but it seems appropriate to end the book with some speculations.

One could argue that the theory of ecological communities is already well suited to incorporating the full spectrum of species and interactions in communities *sensu lato* (see Fig. 2.1). All species interactions result in some form of positive or negative density- or frequency-dependent selection; all species are prone to demographic stochasticity (causing community drift); all species disperse; and new species that arise can be of any kind. We could define community patterns based on vectors of species biomass (a universally applicable abundance metric), including everything from bacteria and arthropods to trees and large mammals. And presto, we have our synthesis of syntheses. But I imagine that many ecologists would not find this satisfactory. Units of plant abundance (e.g., a gram of biomass or an individual tree stem) are sufficiently comparable ecologically to justify placing them together in one abundance vector that is otherwise unstructured (at least initially) according to different ecological "roles." Doing the same for wolves and butterflies and trees and *E. coli* all at once might not be considered useful by many ecologists, although I'd certainly enjoy being convinced otherwise. We'll see.

One could also argue that we already have no shortage of models that simultaneously consider many different kinds of species interactions, and we could consider these models collectively as constituting an overarching theory. However, to my eye, the many hundreds of models of species interactions do not yet comprise much more than the sum of their parts, except perhaps *within* the branches of community ecology defined in Figure 2.1. Especially given the ever-increasing capacity of computers to lend a hand, there is no barrier in principle to building any kind of model we would like, but it's difficult to imagine this path leading to an all-encompassing general theory that simplifies and clarifies our understanding of ecological communities (see also Chap. 4).

There are also some fundamental differences between branches of community ecology on issues as basic as what constitutes a pattern and a process. A food web itself is a description of species interactions (feeding relationships specifically), such that food web "patterns," such as modularity, total length, or connectedness, have built into them what I would consider a low-level process (predation). The same is true of mutualistic networks. In horizontal community ecology, patterns (at least as I consider them in Chap. 2) are characterized based only on which species occur where and when and at what abundance, with interactions among species considered as part of the underlying causes of such patterns. Core questions and hypotheses are thus articulated in very different ways in each of these areas of research, so they are difficult to reconcile within the same conceptual framework.

Finally, all these considerations lead me to think that a grand synthesis—if it ever happens—will not involve fitting one framework or theory into another, or further development of a current line of research. Rather, someone will come at the whole problem from an entirely different angle, which we (or at least I) can't even fathom at the moment. When that time comes, perhaps this book will be considered part of a "problem" that needs fixing. In the meantime, I have attempted to provide a theoretical framework with which to make sense of what I, as a student, found to be a bewildering array of loosely related models, ideas, and concepts. I hope the next generation of students in community ecology will have an easier go of things.

References

Aarssen, L. W. 1997. High productivity in grassland ecosystems: Effected by species diversity or productive species? *Oikos* **80**:183–184.

Adler, P. B., and J. M. Drake. 2008. Environmental variation, stochastic extinction, and competitive coexistence. *American Naturalist* **172**:E186–E195.

Adler, P. B., S. P. Ellner, and J. M. Levine. 2010. Coexistence of perennial plants: An embarrassment of niches. *Ecology Letters* **13**:1019–1029.

Adler, P. B., J. HilleRisLambers, P. C. Kyriakidis, Q. Guan, and J. M. Levine. 2006. Climate variability has a stabilizing effect on the coexistence of prairie grasses. *Proceedings of the National Academy of Sciences USA* **103**:12793–12798.

Adler, P. B., J. HilleRisLambers, and J. M. Levine. 2007. A niche for neutrality. *Ecology Letters* **10**:95–104.

Adler, P. B., E. W. Seabloom, E. T. Borer, H. Hillebrand, Y. Hautier, A. Hector, W. S. Harpole, et al. 2011. Productivity is a poor predictor of plant species richness. *Science* **333**:1750–1753.

Alexander, H. M., B. L. Foster, F. Ballantyne, C. D. Collins, J. Antonovics, and R. D. Holt. 2012. Metapopulations and metacommunities: Combining spatial and temporal perspectives in plant ecology. *Journal of Ecology* **100**:88–103.

Allee, W. E., O. Park, A. E. Emerson, T. Park, and K. P. Schmidt. 1949. *Principles of animal ecology*. W.B. Saunders, Philadelphia, PA.

Allen, A. P., J. H. Brown, and J. F. Gillooly. 2002. Global biodiversity, biochemical kinetics, and the energetic-equivalence rule. *Science* **297**:1545–1548.

Allen, A. P., J. F. Gillooly, V. M. Savage, and J. H. Brown. 2006. Kinetic effects of temperature on rates of genetic divergence and speciation. *Proceedings of the National Academy of Sciences USA* **103**:9130–9135.

Allen, T.F.H., and T. W. Hoekstra. 1992. *Toward a unified ecology*. Columbia University Press, New York.

Allmon, W. D., P. J. Morris, and M. L. McKinney. 1998. An intermediate disturbance hypothesis for maximal speciation. Pages 349–376 *in* M. L. McKinney and J. A. Drake, editors. *Biodiversity dynamics: Turnover of populations, taxa, and communities*. Columbia University Press, New York.

Alroy, J. 2008. Dynamics of origination and extinction in the marine fossil record. *Proceedings of the National Academy of Sciences USA* **105**:11536–11542.

Amarasekare, P. 2000. The geometry of coexistence. *Biological Journal of the Linnean Society* **71**:1–31.

Amarasekare, P. 2003. Competitive coexistence in spatially structured environments: A synthesis. *Ecology Letters* **6**:1109–1122.

Anderson, M. J., T. O. Crist, J. M. Chase, M. Vellend, B. D. Inouye, A. L. Freestone, N. J. Sanders, et al.. 2011. Navigating the multiple meanings of β diversity: A roadmap for the practicing ecologist. *Ecology Letters* **14**:19–28.

Angert, A. L., T. E. Huxman, P. Chesson, and D. L. Venable. 2009. Functional tradeoffs determine species coexistence via the storage effect. *Proceedings of the National Academy of Sciences USA* **106**:11641–11645.

Antonovics, J. 1976. The input from population genetics: "The new ecological genetics." *Systematic Botany* **1**:233–245.

Antonovics, J. 2003. Toward community genomics? *Ecology* **84**:598–601.

Armstrong, R. A., and R. McGehee. 1980. Competitive exclusion. *American Naturalist* **115**:151–170.

Barber, N. A., and R. J. Marquis. 2010. Leaf quality, predators, and stochastic processes in the assembly of a diverse herbivore community. *Ecology* **92**:699–708.

Bascompte, J., and P. Jordano. 2013. *Mutualistic networks*. Princeton University Press, Princeton, NJ.

Beatty, J. 1984. Chance and natural selection. *Philosophy of Science* **51**:183–211.

Beatty, J. 1995. The evolutionary contingency thesis. Pages 45–81 *in* G. Wolters and J. G. Lennox, editors. *Concepts, theories, and rationality in the biological sciences.* University of Pittsburgh Press, Pittsburgh, PA.

Beatty, J. 1997. Why do biologists argue like they do? *Philosophy of Science* **64**:S432–S443.

Beisner, B. E. 2001. Plankton community structure in fluctuating environments and the role of productivity. *Oikos* **95**:496–510.

Beisner, B. E., P. R. Peres-Neto, E. S. Lindström, A. Barnett, and M. L. Longhi. 2006. The role of environmental and spatial processes in structuring lake communities from bacteria to fish. *Ecology* **87**:2985–2991.

Bell, G. 2008. *Selection: The mechanism of evolution.* Oxford University Press, Oxford.

Bell, G., M. Lechowicz, A. Appenzeller, M. Chandler, E. DeBlois, L. Jackson, B. Mackenzie, et al. 1993. The spatial structure of the physical environment. *Oecologia* **96**:114–121.

Belmaker, J., and W. Jetz. 2012. Regional pools and environmental controls of vertebrate richness. *American Naturalist* **179**:512–523.

Belovsky, G. E., D. B. Botkin, T. A. Crowl, K. W. Cummins, J. F. Franklin, M. L. Hunter, A. Joern, et al. 2004. Ten suggestions to strengthen the science of ecology. *BioScience* **54**:345–351.

Bender, E. A., T. J. Case, and M. E. Gilpin. 1984. Perturbation experiments in community ecology: Theory and practice. *Ecology* **65**:1–13.

Bennett, J. A., E. G. Lamb, J. C. Hall, W. M. Cardinal-McTeague, and J. F. Cahill. 2013. Increased competition does not lead to increased phylogenetic overdispersion in a native grassland. *Ecology Letters* **16**:1168–1176.

Bernard-Verdier, M., M.-L. Navas, M. Vellend, C. Violle, A. Fayolle, and E. Garnier.

2012. Community assembly along a soil depth gradient: contrasting patterns of plant trait convergence and divergence in a Mediterranean rangeland. *Journal of Ecology* **100**:1422–1433.

Best, R. J., N. C. Caulk, and J. J. Stachowicz. 2013. Trait vs. phylogenetic diversity as predictors of competition and community composition in herbivorous marine amphipods. *Ecology Letters* **16**:72–80.

Bever, J. D. 2003. Soil community feedback and the coexistence of competitors: Conceptual frameworks and empirical tests. *New Phytologist* **157**:465–473.

Bever, J. D., I. A. Dickie, E. Facelli, J. M. Facelli, J. Klironomos, M. Moora, M. C. Rillig, et al. 2010. Rooting theories of plant community ecology in microbial interactions. *Trends in Ecology & Evolution* **25**:468–478.

Bever, J. D., K. M. Westover, and J. Antonovics. 1997. Incorporating the soil community into plant population dynamics: the utility of the feedback approach. *Journal of Ecology* **85**:561–573.

Borcard, D., and P. Legendre. 2002. All-scale spatial analysis of ecological data by means of principal coordinates of neighbour matrices. *Ecological Modelling* **153**:51–68.

Borcard, D., P. Legendre, and P. Drapeau. 1992. Partialling out the spatial component of ecological variation. *Ecology* **73**:1045–1055.

Borer, E. T., W. S. Harpole, P. B. Adler, E. M. Lind, J. L. Orrock, E. W. Seabloom, and M. D. Smith. 2014. Finding generality in ecology: A model for globally distributed experiments. *Methods in Ecology and Evolution* **5**:65–73.

Borer, E. T., E. W. Seabloom, D. S. Gruner, W. S. Harpole, H. Hillebrand, E. M. Lind, P. B. Adler, et al. 2014. Herbivores and nutrients control grassland plant diversity via light limitation. *Nature* **508**:517–520.

Braun-Blanquet, J. 1932. *Plant sociology: The study of plant communities*. McGraw-Hill, London.

Bray, J. R., and J. T. Curtis. 1957. An ordination of the upland forest communities of southern Wisconsin. *Ecological Monographs* **27**:325–349.

Brown, J. H. 1995. *Macroecology*. University of Chicago Press, Chicago.

Brown, J. H. 2014. Why are there so many species in the tropics? *Journal of Biogeography* **41**:8–22.

Brown, J. H., and A. C. Gibson. 1983. *Biogeography*. C. V. Mosby, St. Louis, MO.

Brown, J. H., J. F. Gillooly, A. P. Allen, V. M. Savage, and G. B. West. 2004. Toward a metabolic theory of ecology. *Ecology* **85**:1771–1789.

Butlin, R., J. Bridle, and D. Schluter. 2009. *Speciation and patterns of diversity*. Cambridge University Press, Cambridge.

Cadotte, M. W. 2006a. Dispersal and species diversity: A meta-analysis. *American Naturalist* **167**:913–924.

Cadotte, M. W. 2006b. Metacommunity influences on community richness at multiple spatial scales: A microcosm experiment. *Ecology* **87**:1008–1016.

Cadotte, M. W., D. V. Mai, S. Jantz, M. D. Collins, M. Keele, and J. A. Drake. 2006. On testing the competition-colonization trade-off in a multispecies assemblage. *American Naturalist* **168**:704–709.

Caley, M. J., and D. Schluter. 1997. The relationship between local and regional diversity. *Ecology* **78**:70–80.

Cardinale, B. J., J. E. Duffy, A. Gonzalez, D. U. Hooper, C. Perrings, P. Venail, A. Narwani, et al. 2012. Biodiversity loss and its impact on humanity. *Nature* **486**:59–67.

Cardinale, B. J., J. P. Wright, M. W. Cadotte, I. T. Carroll, A. Hector, D. S. Srivastava, M. Loreau, et al. 2007. Impacts of plant diversity on biomass production increase through time because of species complementarity. *Proceedings of the National Academy of Sciences USA* **104**:18123–18128.

Carlquist, S. 1967. The biota of long-distance dispersal. V. Plant dispersal to Pacific islands. *Bulletin of the Torrey Botanical Club* **94**:129–162.

Carlquist, S. J. 1974. *Island biology*. Columbia University Press, New York.

Carrara, F., F. Altermatt, I. Rodriguez-Iturbe, and A. Rinaldo. 2012. Dendritic connectivity controls biodiversity patterns in experimental metacommunities. *Proceedings of the National Academy of Sciences USA* **109**:5761–5766.

Carstensen, D. W., J.-P. Lessard, B. G. Holt, M. Krabbe Borregaard, and C. Rahbek. 2013. Introducing the biogeographic species pool. *Ecography* **36**:1310–1318.

Chase, J. M. 2003. Community assembly: When should history matter? *Oecologia* **136**:489–498.

Chase, J. M. 2007. Drought mediates the importance of stochastic community assembly. *Proceedings of the National Academy of Sciences USA* **104**:17430–17434.

Chase, J. M. 2010. Stochastic community assembly causes higher biodiversity in more productive environments. *Science* **328**:1388–1391.

Chase, J. M., E. G. Biro, W. A. Ryberg, and K. G. Smith. 2009. Predators temper the relative importance of stochastic processes in the assembly of prey metacommunities. *Ecology Letters* **12**:1210–1218.

Chase, J. M., and M. A. Leibold. 2002. Spatial scale dictates the productivity-biodiversity relationship. *Nature* **416**:427–430.

Chase, J. M., and M. A. Leibold. 2003. *Ecological niches: Linking classical and contemporary approaches*. University of Chicago Press, Chicago.

Chave, J., H. C. Muller-Landau, and S. A. Levin. 2002. Comparing classical community models: Theoretical consequences for patterns of diversity. *American Naturalist* **159**:1–23.

Chesson, P. 2000a. General theory of competitive coexistence in spatially-varying environments. *Theoretical Population Biology* **58**:211–237.

Chesson, P. 2000b. Mechanisms of maintenance of species diversity. *Annual Review of Ecology and Systematics* **31**:343–366.

Chitty, D. 1957. Self-regulation of numbers through changes in viability. *Cold Spring Harbor Symposia on Quantitative Biology* **22**:277–280.

Clark, J. S. 2009. Beyond neutral science. *Trends in Ecology & Evolution* **24**:8–15.

Clark, J. S. 2010. Individuals and the variation needed for high species diversity in forest trees. *Science* **327**:1129–1132.

Clark, J. S. 2012. The coherence problem with the Unified Neutral Theory of Biodiversity. *Trends in Ecology & Evolution* **27**:198–202.

Clark, J. S., D. Bell, C. Chu, B. Courbaud, M. Dietze, M. Hersh, J. HilleRisLambers, et al. 2010. High-dimensional coexistence based on individual variation: A synthesis of evidence. *Ecological Monographs* **80**:569–608.

Clark, J. S., M. Dietze, S. Chakraborty, P. K. Agarwal, I. Ibanez, S. LaDeau, and M. Wolosin. 2007. Resolving the biodiversity paradox. *Ecology Letters* **10**:647–659.

Clark, J. S., C. Fastie, G. Hurtt, S. T. Jackson, C. Johnson, G. A. King, M. Lewis, et al. 1998. Reid's paradox of rapid plant migration: Dispersal theory and interpretation of paleoecological records. *BioScience* **48**:13–24.

Clark, J. S., S. LaDeau, and I. Ibanez. 2004. Fecundity of trees and the colonization-competition hypothesis. *Ecological Monographs* **74**:415–442.

Clements, F. E. 1916. *Plant succession: An analysis of the development of vegetation.* Carnegie Institute of Washington, Washington, DC.

Clobert, J., M. Baguette, T. G. Benton, J. M. Bullock, and S. Ducatez. 2012. *Dispersal ecology and evolution.* Oxford University Press, Oxford.

Cody, M. L. 1993. Bird diversity components within and between habitats in Australia. Pages 147–158 *in* R. E. Ricklefs and D. Schluter, editors. *Species diversity in ecological communities: Historical and geographic perspectives.* University of Chicago Press, Chicago.

Cody, M. L., and J. M. Diamond. 1975. *Ecology and evolution of communities.* Belknap Press of Harvard University Press, Cambridge, MA.

Comita, L. S., H. C. Muller-Landau, S. Aguilar, and S. P. Hubbell. 2010. Asymmetric density dependence shapes species abundances in a tropical tree community. *Science* **329**:330–332.

Connell, J. H. 1961. The influence of interspecific competition and other factors on the distribution of the barnacle *Chthamalus stellatus*. *Ecology* **42**:710–723.

Connell, J. H. 1970. On the role of natural enemies in preventing competitive exclusion in some marine animals and in rain forest trees. Pages 298–312 *in* P. J. Den Boer and G. R. Gradwell, editors. *Dynamics of populations.* Centre for Agricultural Publishing and Documentation, Wageningen, The Netherlands.

Connell, J. H. 1978. Diversity in tropical rain forests and coral reefs. *Science* **199**:1302–1310.

Connell, J. H. 1983. On the prevalence and relative importance of interspecific competition: evidence from field experiments. *American Naturalist* **122**:661–696.

Connor, E. F., and E. D. McCoy. 1979. The statistics and biology of the species-area relationship. *American Naturalist* **113**:791–833.

Connor, E. F., and D. Simberloff. 1979. The assembly of species communities: chance or competition? *Ecology* **60**:1132–1140.

Cooper, G. J. 2003. *The science of the struggle for existence: On the foundations of ecology.* Cambridge University Press, Cambridge.

Cornell, H. V. 1985. Local and regional richness of cynipine gall wasps on California oaks. *Ecology* **66**:1247–1260.

Cornell, H. V., and S. P. Harrison. 2014. What are species pools and when are they important? *Annual Review of Ecology, Evolution, and Systematics* **45**:45–67.

Cornell, H. V., and J. H. Lawton. 1992. Species interactions, local and regional processes, and limits to the richness of ecological communities: A theoretical perspective. *Journal of Animal Ecology* **61**:1–12.

Cornwell, W. K., and D. D. Ackerly. 2009. Community assembly and shifts in plant trait distributions across an environmental gradient in coastal California. *Ecological Monographs* **79**:109–126.

Cornwell, W. K., D. W. Schwilk, and D. D. Ackerly. 2006. A trait-based test for habitat filtering: Convex hull volume. *Ecology* **87**:1465–1471.

Costello, E. K., K. Stagaman, L. Dethlefsen, B. J. Bohannan, and D. A. Relman. 2012. The application of ecological theory toward an understanding of the human microbiome. *Science* **336**:1255–1262.

Cottenie, K. 2005. Integrating environmental and spatial processes in ecological community dynamics. *Ecology Letters* **8**:1175–1182.

Coyne, J. A.,and H. A. Orr. 2004. *Speciation.* Sinauer Associates, Sunderland, MA.

Crawley, M. 1997. *Plant ecology*, 2nd ed. Blackwell, Oxford.

Csada, R. D., P. C. James, and R.H.M. Espie. 1996. The "file drawer problem" of nonsignificant results: Does it apply to biological research? *Oikos* **76**:591–593.

Currie, D. J. 1991. Energy and large-scale patterns of animal- and plant-species richness. *American Naturalist* **137**:27–49.

Curtis, J. T. 1959. *The vegetation of Wisconsin: An ordination of plant communities.* University of Wisconsin Press, Madison.

D'Avanzo, C. 2008. Symposium 1. Why is ecology hard to learn? *Bulletin of the Ecological Society of America* **89**:462–466.

Damschen, E. I., N. M. Haddad, J. L. Orrock, J. J. Tewksbury, and D. J. Levey. 2006. Corridors increase plant species richness at large scales. *Science* **313**:1284–1286.

Darwin, C. 1859. *On the origin of species.* John Murray, London.

Davies, T. J., V. Savolainen, M. W. Chase, J. Moat, and T. G. Barraclough. 2004. Environmental energy and evolutionary rates in flowering plants. *Proceedings of the Royal Society of London. Series B: Biological Sciences* **271**:2195–2200.

Davis, M. B. 1986. Climatic instability, time lags, and community disequilibrium. Pages 269–284 *in* J. M. Diamond and T. J. Case, editors. *Community ecology.* Harper & Row, New York.

De Cáceres, M., P. Legendre, R. Valencia, M. Cao, L.-W. Chang, G. Chuyong, R. Condit, et al. 2012. The variation of tree beta diversity across a global network of forest plots. *Global Ecology and Biogeography* **21**:1191–1202.

De Frenne, P., F. Rodríguez-Sánchez, D. A. Coomes, L. Baeten, G. Verstraeten, M. Vellend, M. Bernhardt-Römermann, et al. 2013. Microclimate moderates plant responses to macroclimate warming. *Proceedings of the National Academy of Sciences USA* **110**:18561–18565.

Debinski, D. M., and R. D. Holt. 2000. A survey and overview of habitat fragmentation experiments. *Conservation Biology* **14**:342–355.

Desjardins-Proulx, P., and D. Gravel. 2012. A complex speciation–richness relationship in a simple neutral model. *Ecology and Evolution* **2**:1781–1790.

Devictor, V., C. van Swaay, T. Brereton, D. Chamberlain, J. Heliölä, S. Herrando, R. Julliard, et al. 2012. Differences in the climatic debts of birds and butterflies at a continental scale. *Nature Climate Change* **2**:121–124.

Diamond, J. M. 1975. Assembly of species communities. Pages 342–444 *in* M. L. Cody and J. M. Diamond, editors. *Ecology and evolution of communities*. Harvard University Press, Cambridge, MA.

Diamond, J. M. 1986. Overview: laboratory experiments, field experiments, and natural experiments. Pages 3–22 *in* J. M. Diamond and T. J. Case, editors. *Community ecology*. Harper & Row, New York.

Diamond, J. M., and T. J. Case. 1986. *Community ecology*. Harper and Row, New York.

Dornelas, M., S. R. Connolly, and T. P. Hughes. 2006. Coral reef diversity refutes the neutral theory of biodiversity. *Nature* **440**:80–82.

Dornelas, M., N. J. Gotelli, B. McGill, H. Shimadzu, F. Moyes, C. Sievers, and A. E. Magurran. 2014. Assemblage time series reveal biodiversity change but not systematic loss. *Science* **344**:296–299.

Dowle, E. J., M. Morgan-Richards, and S. A. Trewick. 2013. Molecular evolution and the latitudinal biodiversity gradient. *Heredity* **110**:501–510.

Drake, J. A. 1991. Community-assembly mechanics and the structure of an experimental species ensemble. *American Naturalist* **137**:1–26.

Drummond, E. B., and M. Vellend. 2012. Genotypic diversity effects on the performance of *Taraxacum officinale* populations increase with time and environmental favorability. *PLoS One* **7**:e30314.

Dublin, H. T., A. R. E. Sinclair, and J. McGlade. 1990. Elephants and fire as causes of multiple stable states in the Serengeti-Mara woodlands. *Journal of Animal Ecology* **59**:1147–1164.

Dunham, A. E., and S. J. Beaupre. 1998. Ecological experiments: Scale, phenomenology, mechanism and the illusion of generality. Pages 27–49 *in* W. J. Resetarits, Jr. and J. Bernardo, editors. *Experimental ecology: Issues and perspectives*. Oxford University Press, New York.

Dzwonko, Z. 1993. Relations between the floristic composition of isolated young woods and their proximity to ancient woodland. *Journal of Vegetation Science* **4**:693–698.

Edwards, K. F., E. Litchman, and C. A. Klausmeier. 2013. Functional traits explain phytoplankton community structure and seasonal dynamics in a marine ecosystem. *Ecology Letters* **16**:56–63.

Egerton, F. N. 2012. *Roots of ecology: Antiquity to Haeckel*. University of California Press, Berkeley.

Elahi, R., M. I. O'Connor, J. E. Byrnes, J. Dunic, B. K. Eriksson, M. J. Hensel, and P. J. Kearns. 2015. Recent trends in local-scale marine biodiversity reflect community structure and human impacts. *Current Biology* **25**:1938–1943.

Ellner, S., and N. G. Hairston Jr. 1994. Role of overlapping generations in maintaining genetic variation in a fluctuating environment. *American Naturalist* **143**:403–417.

Elton, C. S. 1927. *Animal ecology*. University of Chicago Press, Chicago.

Ernest, S. M., J. H. Brown, K. M. Thibault, E. P. White, and J. R. Goheen. 2008. Zero sum, the niche, and metacommunities: Long-term dynamics of community assembly. *American Naturalist* **172**:E257–E269.

Ewens, W. J. 2004. *Mathematical population genetics. I. Theoretical introduction.* Springer, New York.

Fahrig, L. 2003. Effects of habitat fragmentation on biodiversity. *Annual Review of Ecology, Evolution, and Systematics* **34**:487–515.

Falconer, D. S. and T. F. C. Mackay. 1996. *Introduction to quantitative genetics.* Benjamin Cummings, London.

Fauth, J. E., J. Bernardo, M. Camara, W. J. Resetarits, Jr., J. V. Buskirk, and S. A. McCollum. 1996. Simplifying the jargon of community ecology: A conceptual approach. *American Naturalist* **147**:282–286.

Fauth, J. E., W. J. Resetarits Jr, and H. M. Wilbur. 1990. Interactions between larval salamanders: A case of competitive equality. *Oikos* **58**:91–99.

Fisher, R. A. 1958. *The genetical theory of natural selection.* Dover, New York.

Flinn, K. M., and M. Vellend. 2005. Recovery of forest plant communities in post-agricultural landscapes. *Frontiers in Ecology and the Environment* **3**:243–250.

Flöder, S., J. Urabe, and Z. Kawabata. 2002. The influence of fluctuating light intensities on species composition and diversity of natural phytoplankton communities. *Oecologia* **133**:395–401.

Forbes, A. E., and J. M. Chase. 2002. The role of habitat connectivity and landscape geometry in experimental zooplankton metacommunities. *Oikos* **96**:433–440.

Fox, J. W. 2012. Has any "shortcut" method in ecology ever worked? *Dynamic Ecology.* dynamicecology.wordpress.com/2012/10/23/has-any-shortcut-method-in-ecology-ever-worked.

Fox, J. W. 2013. The intermediate disturbance hypothesis should be abandoned. *Trends in Ecology & Evolution* **28**:86–92.

Fox, J. W., W. A. Nelson, and E. McCauley. 2010. Coexistence mechanisms and the paradox of the plankton: quantifying selection from noisy data. *Ecology* **91**:1774–1786.

Fox, J. W., and D. Srivastava. 2006. Predicting local-regional richness relationships using island biogeography models. *Oikos* **113**:376–382.

Fraser, L. H., H. A. Henry, C. N. Carlyle, S. R. White, C. Beierkuhnlein, J. F. Cahill Jr, B. B. Casper, et al. 2012. Coordinated distributed experiments: an emerging tool for testing global hypotheses in ecology and environmental science. *Frontiers in Ecology and the Environment* **11**:147–155.

Freckleton, R., and A. Watkinson. 2001. Predicting competition coefficients for plant mixtures: Reciprocity, transitivity and correlations with life-history traits. *Ecology Letters* **4**:348–357.

Fukami, T. 2004. Assembly history interacts with ecosystem size to influence species diversity. *Ecology* **85**:3234–3242.

Fukami, T. 2010. Community assembly dynamics in space. Pages 45–54 *in* H. A. Verhoef and P. J. Morin, editors. *Community ecology: Processes, models, and applications.* Oxford University Press, Oxford.

Fukami, T. 2015. Historical contingency in community assembly: integrating niches, species pools, and priority effects. *Annual Review of Ecology, Evolution, and Systematics* **46**:1–23.

Fukami, T., and M. Nakajima. 2011. Community assembly: Alternative stable states or alternative transient states? *Ecology Letters* **14**:973–984.

Fussmann, G., M. Loreau, and P. Abrams. 2007. Eco-evolutionary dynamics of communities and ecosystems. *Functional Ecology* **21**:465–477.

Gaston, K., and T. Blackburn. 2000. *Pattern and process in macroecology*. Blackwell, Oxford.

Gause, G. F. 1934. *The struggle for existence*. Williams and Wilkins, Baltimore.

Gewin, V. 2006. Beyond neutrality—ecology finds its niche. *PLoS Biology* **4**:e278.

Gigerenzer, G., Z. Swijtink, T. Porter, L. Daston, J. Beatty, and L. Krüger. 1989. *Empire of chance: How probability changed science and everyday life*. Cambridge University Press, Cambridge.

Gilbert, B., and J. R. Bennett. 2010. Partitioning variation in ecological communities: Do the numbers add up? *Journal of Applied Ecology* **47**:1071–1082.

Gilbert, B., and M. J. Lechowicz. 2004. Neutrality, niches, and dispersal in a temperate forest understory. *Proceedings of the National Academy of Sciences USA* **101**:7651–7656.

Gilbert, F., A. Gonzalez, and I. Evans-Freke. 1998. Corridors maintain species richness in the fragmented landscapes of a microecosystem. *Proceedings of the Royal Society of London. Series B: Biological Sciences* **265**:577–582.

Gillespie, R. 2004. Community assembly through adaptive radiation in Hawaiian spiders. *Science* **303**:356–359.

Gillman, L. N., P. McBride, D. J. Keeling, H. A. Ross, and S. D. Wright. 2011. Are rates of molecular evolution in mammals substantially accelerated in warmer environments? Reply. *Proceedings of the Royal Society B: Biological Sciences* **278**:1294–1297.

Gillman, L. N., and S. D. Wright. 2014. Species richness and evolutionary speed: The influence of temperature, water and area. *Journal of Biogeography* **41**:39–51.

Gilpin, M. E. 1975. Limit cycles in competition communities. *American Naturalist* **109**:51–60.

Gleason, H. A. 1926. The individualistic concept of the plant association. *Bulletin of the Torrey Botanical Club* **53**:7–26.

Godoy, O., N. J. B. Kraft, and J. M. Levine. 2014. Phylogenetic relatedness and the determinants of competitive outcomes. *Ecology Letters* **17**:836–844.

Goldberg, D. E., and A. M. Barton. 1992. Patterns and consequences of interspecific competition in natural communities: A review of field experiments with plants. *American Naturalist* **139**:771–801.

Gonzalez, A., and E. J. Chaneton. 2002. Heterotroph species extinction, abundance and biomass dynamics in an experimentally fragmented microecosystem. *Journal of Animal Ecology* **71**:594–602.

Gotelli, N. J., and R. K. Colwell. 2001. Quantifying biodiversity: Procedures and pitfalls in the measurement and comparison of species richness. *Ecology Letters* **4**:379–391.

Gotelli, N. J., and G. R. Graves. 1996. *Null models in ecology*. Smithsonian Institution Press, Washington, DC.

Gotelli, N. J., and D. J. McCabe. 2002. Species co-occurrence: A meta-analysis of J. M. Diamond's assembly rules model. *Ecology* **83**:2091–2096.

Grace, J. B., S. Harrison, and E. I. Damschen. 2011. Local richness along gradients in the Siskiyou herb flora: R. H. Whittaker revisited. *Ecology* **92**:108–120.

Graham, M. H., and P. K. Dayton. 2002. On the evolution of ecological ideas: Paradigms and scientific progress. *Ecology* **83**:1481–1489.

Grant, P. R., and B. R. Grant. 2002. Unpredictable evolution in a 30-year study of Darwin's finches. *Science* **296**:707–711.

Gravel, D., C. D. Canham, M. Beaudet, and C. Messier. 2006. Reconciling niche and neutrality: The continuum hypothesis. *Ecology Letters* **9**:399–409.

Green, P. T., K. E. Harms, and J. H. Connell. 2014. Nonrandom, diversifying processes are disproportionately strong in the smallest size classes of a tropical forest. *Proceedings of the National Academy of Sciences USA* **111**:18649–18654.

Greene, D., and E. Johnson. 1994. Estimating the mean annual seed production of trees. *Ecology* **75**:642–647.

Grime, J. P. 1973. Competitive exclusion in herbaceous vegetation. *Nature* **242**:344–347.

Grime, J. P. 1979. *Plant strategies and vegetation processes*. Wiley, London.

Grime, J. P. 2006. *Plant strategies, vegetation processes, and ecosystem properties*. Wiley, London.

Gurevitch, J., L. L. Morrow, A. Wallace, and J. S. Walsh. 1992. A meta-analysis of competition in field experiments. *American Naturalist* **140**:539–572.

Gurevitch, J., S. M. Scheiner, and G. A. Fox. 2006. *The ecology of plants*, 2nd ed. Sinauer Associates, Sunderland, MA.

Haegeman, B., and M. Loreau. 2011. A mathematical synthesis of niche and neutral theories in community ecology. *Journal of Theoretical Biology* **269**:150–165.

Haegeman, B., and M. Loreau. 2014. General relationships between consumer dispersal, resource dispersal and metacommunity diversity. *Ecology Letters* **17**:175–184.

Hairston, N. G. 1989. *Ecological experiments: Purpose, design and execution*. Cambridge University Press, Cambridge.

Hájek, M., J. Roleček, K. Cottenie, K. Kintrová, M. Horsák, A. Poulíčková, P. Hájková, et al. 2011. Environmental and spatial controls of biotic assemblages in a discrete semi-terrestrial habitat: Comparison of organisms with different dispersal abilities sampled in the same plots. *Journal of Biogeography* **38**:1683–1693.

Hansen, S. K., P. B. Rainey, J. A. Haagensen, and S. Molin. 2007. Evolution of species interactions in a biofilm community. *Nature* **445**:533–536.

Harmon-Threatt, A. N., and D. D. Ackerly. 2013. Filtering across spatial scales: Phylogeny, biogeography and community structure in bumble bees. *PLoS One* **8**:e60446.

Harms, K. E., S. J. Wright, O. Calderon, A. Hernandez, and E. A. Herre. 2000. Pervasive density-dependent recruitment enhances seedling diversity in a tropical forest. *Nature* **404**:493–495.

Harper, J. L. 1977. *Population biology of plants*. Blackburn Press, Caldwell, NJ.

Harrison, S. 1999. Local and regional diversity in a patchy landscape: Native, alien, and endemic herbs on serpentine. *Ecology* **80**:70–80.

Harrison, S., and E. Bruna. 1999. Habitat fragmentation and large-scale conservation: What do we know for sure? *Ecography* **22**:225–232.

Harrison, S., H. D. Safford, J. B. Grace, J. H. Viers, and K. F. Davies. 2006. Regional and local species richness in an insular environment: Serpentine plants in California. *Ecological Monographs* **76**:41–56.

Harte, J. 2011. *Maximum entropy and ecology: A theory of abundance, distribution, and energetics*. Oxford University Press, Oxford.

Harte, J., and E. A. Newman. 2014. Maximum information entropy: A foundation for ecological theory. *Trends in Ecology & Evolution* **29**:384–389.

Harte, J., T. Zillio, E. Conlisk, and A. B. Smith. 2008. Maximum entropy and the state-variable approach to macroecology. *Ecology* **89**:2700–2711.

Hartl, D. L., and A. G. Clark. 1997. *Principles of population genetics*. Sinauer Associates, Sunderland, MA.

Hastings, A. 2004. Transients: The key to long-term ecological understanding? *Trends in Ecology & Evolution* **19**:39–45.

Hawkins, B. A., J. A. F. Diniz-Filho, C. A. Jaramillo, and S. A. Soeller. 2007. Climate, niche conservatism, and the global bird diversity gradient. *American Naturalist* **170**:S16–S27.

Hawkins, B. A., R. Field, H. V. Cornell, D. J. Currie, J.-F. Guégan, D. M. Kaufman, J. T. Kerr, et al. 2003. Energy, water, and broad-scale geographic patterns of species richness. *Ecology* **84**:3105–3117.

Helmus, M. R., D. L. Mahler, and J. B. Losos. 2014. Island biogeography of the Anthropocene. *Nature* **513**:543–546.

Hendry, A. P. *Eco-evolutionary dynamics*. Princeton University Press, Princeton, NJ, forthcoming.

HilleRisLambers, J., P. B. Adler, W. S. Harpole, J. M. Levine, and M. M. Mayfield. 2012. Rethinking community assembly through the lens of coexistence theory. *Annual Review of Ecology, Evolution, and Systematics* **43**:227–248.

Hirota, M., M. Holmgren, E. H. Van Nes, and M. Scheffer. 2011. Global resilience of tropical forest and savanna to critical transitions. *Science* **334**:232–235.

Hodgson, J. G., P. J. Wilson, R. Hunt, J. P. Grime, and K. Thompson. 1999. Allocating C-S-R plant functional types: A soft approach to a hard problem. *Oikos* **85**:282–294.

Holt, R. D. 1977. Predation, apparent competition, and the structure of prey communities. *Theoretical Population Biology* **12**:197–229.

Holt, R. D. 1993. Ecology at the mesoscale: The influence of regional processes on local communities. Pages 77–88 *in* R. E. Ricklefs and D. Schluter, editors. *Species diversity in ecological communities: Historical and geographic perspectives*. University of Chicago Press, Chicago.

Holt, R. D. 1997. Community modules. Pages 333–349 *in* A. C. Gange and V. K. Brown, editors. *Multitrophic interactions in terrestrial ecosystems*. Blackwell Science, London.

Holt, R. D. 2005. On the integration of community ecology and evolutionary biology: Historical perspectives and current prospects. Pages 235–271 *in* K. Cuddington and B. Beisner, editors. *Ecological paradigms lost: Routes of theory change.* Elsevier, London.

Holt, R. D., J. Grover, and D. Tilman. 1994. Simple rules for interspecific dominance in systems with exploitative and apparent competition. *American Naturalist* **144**:741–771.

Holyoak, M., M. A. Leibold, and R. D. Holt. 2005. *Metacommunities: spatial dynamics and ecological communities.* University of Chicago Press, Chicago.

Holyoak, M., and M. Loreau. 2006. Reconciling empirical ecology with neutral community models. *Ecology* **87**:1370–1377.

Howeth, J. G., and M. A. Leibold. 2010. Species dispersal rates alter diversity and ecosystem stability in pond metacommunities. *Ecology* **91**:2727–2741.

Hoyle, M., and F. Gilbert. 2004. Species richness of moss landscapes unaffected by short-term fragmentation. *Oikos* **105**:359–367.

Hu, X. S., F. He, and S. P. Hubbell. 2006. Neutral theory in macroecology and population genetics. *Oikos* **113**:548–556.

Hubbell, S. 2009. Neutral theory and the theory of island biogeography. Pages 240–261 *in* J. B. Losos and R. E. Ricklefs, editors. *The theory of island biogeography revisited.* Princeton University Press, Princeton, NJ.

Hubbell, S. P. 2001. *The unified neutral theory of biogeography and biodiversity.* Princeton University Press, Princeton, NJ.

Hubbell, S. P. 2006. Neutral theory and the evolution of ecological equivalence. *Ecology* **87**:1387–1398.

Hubbell, S. P., and R. B. Foster. 1986. Biology, chance, and history and the structure of tropical rain forest tree communities. Pages 314–330 *in* J. M. Diamond and T. J. Case, editors. *Community ecology.* Harper & Row, New York.

Hughes, A. R., J. E. Byrnes, D. L. Kimbro, and J. J. Stachowicz. 2007. Reciprocal relationships and potential feedbacks between biodiversity and disturbance. *Ecology Letters* **10**:849–864.

Hughes, A. R., B. D. Inouye, M. T. Johnson, N. Underwood, and M. Vellend. 2008. Ecological consequences of genetic diversity. *Ecology Letters* **11**:609–623.

Hughes, T. P. 1994. Catastrophes, phase shifts, and large-scale degradation of a Caribbean coral reef. *Science* **265**:1547–1547.

Huisman, J., and F. J. Weissing. 1999. Biodiversity of plankton by species oscillations and chaos. *Nature* **402**:407–410.

Huisman, J., and F. J. Weissing. 2001. Fundamental unpredictability in multispecies competition. *American Naturalist* **157**:488–494.

Huston, M. 1979. A general hypothesis of species diversity. *American Naturalist* **113**:81–101.

Huston, M. A. 1994. *Biological diversity: The coexistence of species on changing landscapes.* Cambridge University Press, Cambridge.

Huston, M. A. 2014. Disturbance, productivity, and species diversity: Empiricism versus logic in ecological theory. *Ecology* **95**:2382–2396.

Hutchinson, G. E. 1959. Homage to Santa Rosalia or why are there so many kinds of animals? *American Naturalist* **93**:145–159.

Hutchinson, G. E. 1961. The paradox of the plankton. *American Naturalist* **95**:137–145.

Isbell, F., D. Tilman, S. Polasky, S. Binder, and P. Hawthorne. 2013. Low biodiversity state persists two decades after cessation of nutrient enrichment. *Ecology Letters* **16**:454–460.

Jablonski, D., K. Roy, and J. W. Valentine. 2006. Out of the tropics: Evolutionary dynamics of the latitudinal diversity gradient. *Science* **314**:102–106.

Jackson, S. T., and J. L. Blois. 2015. Community ecology in a changing environment: Perspectives from the Quaternary. *Proceedings of the National Academy of Sciences USA* **112**:4915–4921.

Jacobson, B., and P. R. Peres-Neto. 2010. Quantifying and disentangling dispersal in metacommunities: How close have we come? How far is there to go? *Landscape Ecology* **25**:495–507.

Jacquemyn, H., J. Butaye, M. Dumortier, M. Hermy, and N. Lust. 2001. Effects of age and distance on the composition of mixed deciduous forest fragments in an agricultural landscape. *Journal of Vegetation Science* **12**:635–642.

Jakobsson, A., and O. Eriksson. 2003. Trade-offs between dispersal and competitive ability: A comparative study of wind-dispersed Asteraceae forbs. *Evolutionary Ecology* **17**:233–246.

Janzen, D. H. 1970. Herbivores and the number of tree species in tropical forests. *American Naturalist* **104**:501–528.

Jetz, W., and P. V. Fine. 2012. Global gradients in vertebrate diversity predicted by historical area-productivity dynamics and contemporary environment. *PLoS Biology* **10**:e1001292.

John, R., J. W. Dalling, K. E. Harms, J. B. Yavitt, R. F. Stallard, M. Mirabello, S. P. Hubbell, et al. 2007. Soil nutrients influence spatial distributions of tropical tree species. *Proceedings of the National Academy of Sciences USA* **104**:864–869.

Jolliffe, P. A. 2000. The replacement series. *Journal of Ecology* **88**:371–385.

Kadmon, R. 1995. Nested species subsets and geographic isolation: A case study. *Ecology* **76**:458–465.

Kadmon, R., and H. R. Pulliam. 1993. Island biogeography: Effect of geographical isolation on species composition. *Ecology* **74**:978–981.

Kalmar, A., and D. J. Currie. 2006. A global model of island biogeography. *Global Ecology and Biogeography* **15**:72–81.

Kareiva, P. 1994. Special feature: Space; The final frontier for ecological theory. *Ecology* **75**:1–1.

Kassen, R. 2014. *Experimental evolution and the nature of biodiversity*. Roberts & Company, Greenwood Village, CO.

Keddy, P. A. 2001. *Competition*. Springer, New York.

Kerr, B., M. A. Riley, M. W. Feldman, and B. J. Bohannan. 2002. Local dispersal promotes biodiversity in a real-life game of rock–paper–scissors. *Nature* **418**:171–174.

Kettlewell, H. 1961. The phenomenon of industrial melanism in Lepidoptera. *Annual Review of Entomology* **6**:245–262.

Kimura, M. 1962. On the probability of fixation of mutant genes in a population. *Genetics* **47**:713.

Kingsland, S. E. 1995. *Modeling nature: Episodes in the history of population ecology.* University of Chicago Press, Chicago.

Knapp, A. K., and C. D'Avanzo. 2010. Teaching with principles: Toward more effective pedagogy in ecology. *Ecosphere* **1**:art15. doi:10.1890/ES10-00013.1.

Kneitel, J. M., and J. M. Chase. 2004. Trade-offs in community ecology: Linking spatial scales and species coexistence. *Ecology Letters* **7**:69–80.

Kneitel, J. M., and T. E. Miller. 2003. Dispersal rates affect species composition in metacommunities of *Sarracenia purpurea* inquilines. *American Naturalist* **162**:165–171.

Kolasa, J., and C. D. Rollo. 1991. Heterogeneity of heterogeneity. Pages 1–23 *in* J. Kolasa and S.T.A. Pickett, editors. *Ecological heterogeneity.* Springer, New York.

Kozak, K. H., and J. J. Wiens. 2012. Phylogeny, ecology, and the origins of climate-richness relationships. *Ecology* **93**:S167–S181.

Kraft, N.J.B., and D. D. Ackerly. 2010. Functional trait and phylogenetic tests of community assembly across spatial scales in an Amazonian forest. *Ecological Monographs* **80**:401–422.

Kraft, N.J.B., O. Godoy, and J. M. Levine. 2015. Plant functional traits and the multidimensional nature of species coexistence. *Proceedings of the National Academy of Sciences USA* **112**:797–802.

Kraft, N.J.B., R. Valencia, and D. D. Ackerly. 2008. Functional traits and niche-based tree community assembly in an Amazonian forest. *Science* **322**:580–582.

Kraft, N.J.B., P. B. Adler, O. Godoy, E. C. James, S. Fuller, and J. M. Levine. 2015. Community assembly, coexistence and the environmental filtering metaphor. *Functional Ecology* **29**:592–599.

Krebs, C. J. 2009. *Ecology: The experimental analysis of distribution and abundance*, 6th ed. Pearson, Upper Saddle River, NJ.

Krug, A. Z., D. Jablonski, and J. W. Valentine. 2007. Contrarian clade confirms the ubiquity of spatial origination patterns in the production of latitudinal diversity gradients. *Proceedings of the National Academy of Sciences USA* **104**:18129–18134.

Krug, A. Z., D. Jablonski, J. W. Valentine, and K. Roy. 2009. Generation of Earth's first-order biodiversity pattern. *Astrobiology* **9**:113–124.

Kutschera, U., and K. Niklas. 2004. The modern theory of biological evolution: An expanded synthesis. *Naturwissenschaften* **91**:255–276.

Laanisto, L., R. Tamme, I. Hiiesalu, R. Szava-Kovats, A. Gazol, and M. Pärtel. 2013. Microfragmentation concept explains non-positive environmental heterogeneity-diversity relationships. *Oecologia* **171**:217–226.

Lacourse, T. 2009. Environmental change controls postglacial forest dynamics through interspecific differences in life-history traits. *Ecology* **90**:2149–2160.

Laland, K., T. Uller, M. Feldman, K. Sterelny, G. B. Müller, A. Moczek, E. Jablonka, et al. 2014. Does evolutionary theory need a rethink? *Nature* **514**:161.

Laliberté, E., and P. Legendre. 2010. A distance-based framework for measuring functional diversity from multiple traits. *Ecology* **91**:299–305.

Laliberté, E., G. Zemunik, and B. L. Turner. 2014. Environmental filtering explains variation in plant diversity along resource gradients. *Science* **345**:1602–1605.

Laurance, W. F., T. E. Lovejoy, H. L. Vasconcelos, E. M. Bruna, R. K. Didham, P. C. Stouffer, C. Gascon, et al. 2002. Ecosystem decay of Amazonian forest fragments: A 22-year investigation. *Conservation Biology* **16**:605–618.

Lawton, J. H. 1991. Warbling in different ways. *Oikos* **60**:273–274.

Lawton, J. H. 1999. Are there general laws in ecology? *Oikos* **84**:177–192.

Lee, S. C. 2006. Habitat complexity and consumer-mediated positive feedbacks on a Caribbean coral reef. *Oikos* **112**:442–447.

Legendre, P., D. Borcard, and P. R. Peres-Neto. 2005. Analyzing beta diversity: Partitioning the spatial variation of community composition data. *Ecological Monographs* **75**:435–450.

Legendre, P., and M. J. Fortin. 1989. Spatial pattern and ecological analysis. *Vegetatio* **80**:107–138.

Legendre, P., and L.F.J. Legendre. 2012. *Numerical ecology*, 3rd ed. Elsevier Science, The Netherlands.

Leibold, M. A., M. Holyoak, N. Mouquet, P. Amarasekare, J. Chase, M. Hoopes, R. Holt, et al. 2004. The metacommunity concept: A framework for multi-scale community ecology. *Ecology Letters* **7**:601–613.

Leibold, M. A., and M. A. McPeek. 2006. Coexistence of the niche and neutral perspectives in community ecology. *Ecology* **87**:1399–1410.

Leishman, M. R. 2001. Does the seed size/number trade-off model determine plant community structure? An assessment of the model mechanisms and their generality. *Oikos* **93**:294–302.

Lerner, I. M., and E. R. Dempster. 1962. Indeterminism in interspecific competition. *Proceedings of the National Academy of Sciences USA* **48**:821.

Lessard, J.-P., J. Belmaker, J. A. Myers, J. M. Chase, and C. Rahbek. 2012. Inferring local ecological processes amid species pool influences. *Trends in Ecology & Evolution* **27**:600–607.

Letcher, S. G. 2010. Phylogenetic structure of angiosperm communities during tropical forest succession. *Proceedings of the Royal Society B: Biological Sciences* **277**:97–104.

Levene, H. 1953. Genetic equilibrium when more than one ecological niche is available. *American Naturalist* **87**:331–333.

Levin, S. A. 1972. A mathematical analysis of the genetic feedback mechanism. *American Naturalist* **106**:145–164.

Levin, S. A. 1992. The problem of pattern and scale in ecology: The Robert H. MacArthur award lecture. *Ecology* **73**:1943–1967.

Levin, S. A. 1998. Ecosystems and the biosphere as complex adaptive systems. *Ecosystems* **1**:431–436.

Levine, J. M., P. B. Adler, and J. HilleRisLambers. 2008. On testing the role of niche differences in stabilizing coexistence. *Functional Ecology* **22**:934–936.

Levine, J. M., and J. HilleRisLambers. 2009. The importance of niches for the maintenance of species diversity. *Nature* **461**:254–257.

Levine, J. M., and D. J. Murrell. 2003. The community-level consequences of seed dispersal patterns. *Annual Review of Ecology, Evolution, and Systematics* **34**:549–574.

Levine, J. M., and M. Rees. 2002. Coexistence and relative abundance in annual plant assemblages: The roles of competition and colonization. *American Naturalist* **160**:452–467.

Levins, R., and D. Culver. 1971. Regional coexistence of species and competition between rare species. *Proceedings of the National Academy of Sciences USA* **68**:1246–1248.

Levins, R., and R. Lewontin. 1980. Dialectics and reductionism in ecology. *Synthese* **43**:47–78.

Lewontin, R. C. 1969. The meaning of stability. *Brookhaven Symposia in Biology* **22**:13–23.

Lewontin, R. C. 1970. The units of selection. *Annual Review of Ecology and Systematics* **1**:1–18.

Lewontin, R. C. 1974. *The genetic basis of evolutionary change*. Columbia University Press, New York.

Lewontin, R. C. 2004. The problems of population genetics. Pages 5–23 *in* R. S. Singh and C. B. Krimbas, editors. *Evolutionary genetics: From molecules to morphology*. Cambridge University Press, Cambridge.

Lilley, P. L., and M. Vellend. 2009. Negative native–exotic diversity relationship in oak savannas explained by human influence and climate. *Oikos* **118**:1373–1382.

Litchman, E., and C. A. Klausmeier. 2008. Trait-based community ecology of phytoplankton. *Annual Review of Ecology, Evolution, and Systematics* **39**:615–639.

Logue, J. B., N. Mouquet, H. Peter, and H. Hillebrand. 2011. Empirical approaches to metacommunities: A review and comparison with theory. *Trends in Ecology & Evolution* **26**:482–491.

Lomolino, M. V. 1982. Species-area and species-distance relationships of terrestrial mammals in the Thousand Island Region. *Oecologia* **54**:72–75.

Lomolino, M. V., B. R. Riddle, R. J. Whittaker, and J. H. Brown. 2010. *Biogeography*, 4th ed. Sinauer Associates, Sunderland, MA.

Loreau, M. 2010. *From populations to ecosystems: Theoretical foundations for a new ecological synthesis*. Princeton University Press, Princeton, NJ.

Loreau, M. and A. Hector. 2001. Partitioning selection and complementarity in biodiversity experiments. *Nature* **412**:72–76.

Loreau, M., and N. Mouquet. 1999. Immigration and the maintenance of local species diversity. *American Naturalist* **154**:427–440.

Losos, J. B., and C. E. Parent. 2009. The speciation-area relationship. Pages 361–378 *in* J. B. Losos and R. E. Ricklefs, editors. *The theory of island biogeography revisited*. Princeton University Press, Princeton, NJ.

Losos, J. B., and R. E. Ricklefs. 2009. *The theory of island biogeography revisited.* Princeton University Press, Princeton, NJ.

Losos, J. B., and D. Schluter. 2000. Analysis of an evolutionary species-area relationship. *Nature* **408**:847–850.

Lowe, W. H., and M. A. McPeek. 2014. Is dispersal neutral? *Trends in Ecology & Evolution* **29**:444–450.

Lundholm, J. T. 2009. Plant species diversity and environmental heterogeneity: Spatial scale and competing hypotheses. *Journal of Vegetation Science* **20**:377–391.

Lundholm, J. T., and D. W. Larson. 2003. Temporal variability in water supply controls seedling diversity in limestone pavement microcosms. *Journal of Ecology* **91**:966–975.

MacArthur, R. H. 1958. Population ecology of some warblers of northeastern coniferous forests. *Ecology* **39**:599–619.

MacArthur, R. H. 1964. Environmental factors affecting bird species diversity. *American Naturalist* **98**:387–397.

MacArthur, R. H. 1969. Patterns of communities in the tropics. *Biological Journal of the Linnean Society* **1**:19–30.

MacArthur, R. H. 1972. *Geographical ecology: Patterns in the distribution of species.* Princeton University Press, Princeton, NJ.

MacArthur, R. H., and J. W. MacArthur. 1961. On bird species diversity. *Ecology* **42**:594–598.

MacArthur, R. H., and E. O. Wilson. 1967. *The theory of island biogeography.* Princeton University Press, Princeton, NJ.

MacDonald, G. M., K. D. Bennett, S. T. Jackson, L. Parducci, F. A. Smith, J. P. Smol, and K. J. Willis. 2008. Impacts of climate change on species, populations and communities: Palaeobiogeographical insights and frontiers. *Progress in Physical Geography* **32**:139–172.

MacDougall, A. S., J. R. Bennett, J. Firn, E. W. Seabloom, E. T. Borer, E. M. Lind, J. L. Orrock, et al. 2014. Anthropogenic-based regional-scale factors most consistently explain plot-level exotic diversity in grasslands. *Global Ecology and Biogeography* **23**:802–810.

MacDougall, A. S., B. Gilbert, and J. M. Levine. 2009. Plant invasions and the niche. *Journal of Ecology* **97**:609–615.

Mack, M. C., C. M. D'Antonio, and R. E. Ley. 2001. Alteration of ecosystem nitrogen dynamics by exotic plants: A case study of C_4 grasses in Hawaii. *Ecological Applications* **11**:1323–1335.

Mackey, R. L., and D. J. Currie. 2001. The diversity-disturbance relationship: Is it generally strong and peaked? *Ecology* **82**:3479–3492.

Magurran, A. E., and R. M. May. 1999. *Evolution of biological diversity.* Oxford University Press, Oxford.

Magurran, A. E., and B. J. McGill. 2010. *Biological diversity: Frontiers in measurement and assessment.* Oxford University Press, Oxford.

Marcotte, G., and M. M. Grandtner. 1974. Étude écologique de la végétation forestière du Mont Mégantic. Gouvernement du Québec, Québec.

Margalef, R. 1978. Life-forms of phytoplankton as survival alternatives in an unstable environment. *Oceanologica Acta* **1**:493–509.

Marquet, P. A., A. P. Allen, J. H. Brown, J. A. Dunne, B. J. Enquist, J. F. Gillooly, P. A. Gowaty, et al. 2014. On theory in ecology. *BioScience* **64**:701–710.

Martin, P. R. 2014. Trade-offs and biological diversity: integrative answers to ecological questions. Pages 291–308 *in* L. B. Martin, C. K. Ghalambor, and H. A. Woods, editors. *Integrative organismal biology*. Wiley, New York.

Maurer, B. A. 1999. *Untangling ecological complexity: The macroscopic perspective*. University of Chicago Press, Chicago.

May, R. M. 1974. Biological populations with nonoverlapping generations: Stable points, stable cycles, and chaos. *Science* **186**:645–647.

May, R. M. 1976. *Theoretical ecology: Principles and applications*. W.B. Saunders, Philadelphia.

Mayfield, M. M., and J. M. Levine. 2010. Opposing effects of competitive exclusion on the phylogenetic structure of communities. *Ecology Letters* **13**:1085–1093.

Mayr, E. 1982. *The growth of biological thought: Diversity, evolution, and inheritance*. Belknap Press of Harvard University Press, Cambridge, MA.

McCann, K. S. 2011. *Food webs*. Princeton University Press, Princeton, NJ.

McGill, B. 2003a. Strong and weak tests of macroecological theory. *Oikos* **102**:679–685.

McGill, B. J. 2003b. A test of the unified neutral theory of biodiversity. *Nature* **422**:881–885.

McGill, B. J., M. Dornelas, N. J. Gotelli, and A. E. Magurran. 2015. Fifteen forms of biodiversity trend in the Anthropocene. *Trends in Ecology & Evolution* **30**:104–113.

McGill, B. J., B. J. Enquist, E. Weiher, and M. Westoby. 2006. Rebuilding community ecology from functional traits. *Trends in Ecology & Evolution* **21**:178–185.

McGill, B. J., R. S. Etienne, J. S. Gray, D. Alonso, M. J. Anderson, H. K. Benecha, M. Dornelas, et al. 2007. Species abundance distributions: moving beyond single prediction theories to integration within an ecological framework. *Ecology Letters* **10**:995–1015.

McGill, B. J., and J. C. Nekola. 2010. Mechanisms in macroecology: AWOL or purloined letter? Towards a pragmatic view of mechanism. *Oikos* **119**:591–603.

McIntosh, R. P. 1980. The background and some current problems of theoretical ecology. *Synthese* **43**:195–255.

McIntosh, R. P. 1985. *The background of ecology: Concept and theory*. Cambridge University Press, Cambridge.

McIntosh, R. P. 1987. Pluralism in ecology. *Annual Review of Ecology and Systematics* **18**:321–341.

McKinney, M. L., and J. A. Drake. 1998. *Biodiversity dynamics: Turnover of populations, taxa, and communities*. Columbia University Press, New York.

McLachlan, J. S., J. S. Clark, and P. S. Manos. 2005. Molecular indicators of tree migration capacity under rapid climate change. *Ecology* **86**:2088–2098.

McPeek, M. A. 2007. The macroevolutionary consequences of ecological differences among species. *Palaeontology* **50**:111–129.

McShea, D. W., and R. N. Brandon. 2010. *Biology's first law: The tendency for diversity and complexity to increase in evolutionary systems*. University of Chicago Press, Chicago.

Meijer, M. 2000. *Biomanipulation in the Netherlands: 15 years of experience*. Wageningen University, Wageningen, The Netherlands.

Menezes, S., D. J. Baird, and A.M.V.M. Soares. 2010. Beyond taxonomy: A review of macroinvertebrate trait-based community descriptors as tools for freshwater biomonitoring. *Journal of Applied Ecology* **47**:711–719.

Merriam, C. H. 1894. Laws of temperature control of the geographic distribution of terrestrial animals and plants. *National Geographic* **6**:229–238.

Mertz, D. B., D. Cawthon, and T. Park. 1976. An experimental analysis of competitive indeterminacy in *Tribolium*. *Proceedings of the National Academy of Sciences USA* **73**:1368–1372.

Mesoudi, A. 2011. *Cultural evolution: How Darwinian theory can explain human culture and synthesize the social sciences*. University of Chicago Press, Chicago.

Mittelbach, G. G. 2012. *Community ecology*. Sinauer Associates, Sunderland, MA.

Mittelbach, G. G., D. W. Schemske, H. V. Cornell, A. P. Allen, J. M. Brown, M. B. Bush, S. P. Harrison, et al. 2007. Evolution and the latitudinal diversity gradient: Speciation, extinction and biogeography. *Ecology Letters* **10**:315–331.

Molofsky, J., R. Durrett, J. Dushoff, D. Griffeath, and S. Levin. 1999. Local frequency dependence and global coexistence. *Theoretical Population Biology* **55**:270–282.

Montaña, C. G., K. O. Winemiller, and A. Sutton. 2013. Intercontinental comparison of fish ecomorphology: Null model tests of community assembly at the patch scale in rivers. *Ecological Monographs* **84**:91–107.

Moran, P.A.P. 1958. Random processes in genetics. Pages 60–71 *in Mathematical Proceedings of the Cambridge Philosophical Society*. Cambridge University Press, Cambridge.

Morin, P. J. 2011. *Community ecology*. Wiley, New York.

Mouquet, N., and M. Loreau. 2003. Community patterns in source-sink metacommunities. *American Naturalist* **162**:544–557.

Mumby, P. J. 2009. Phase shifts and the stability of macroalgal communities on Caribbean coral reefs. *Coral Reefs* **28**:761–773.

Mumby, P. J., A. Hastings, and H. J. Edwards. 2007. Thresholds and the resilience of Caribbean coral reefs. *Nature* **450**:98–101.

Munday, P. L. 2004. Competitive coexistence of coral-dwelling fishes: The lottery hypothesis revisited. *Ecology* **85**:623–628.

Murdoch, W. W., C. J. Briggs, and R. M. Nisbet. 2013. *Consumer-resource dynamics*. Princeton University Press, Princeton, NJ.

Myers, J. A., and K. E. Harms. 2009. Seed arrival, ecological filters, and plant species richness: A meta-analysis. *Ecology Letters* **12**:1250–1260.

Naeem, S. 2001. Experimental validity and ecological scale as criteria for evaluating research programs. Pages 223–250 *in* R. H. Gardner, W. M. Kemp, V. S. Kennedy, and J. E. Petersen, editors. *Scaling relations in experimental ecology*. Columbia University Press, New York.

Narwani, A., M. A. Alexandrou, T. H. Oakley, I. T. Carroll, and B. J. Cardinale. 2013. Experimental evidence that evolutionary relatedness does not affect the ecological mechanisms of coexistence in freshwater green algae. *Ecology Letters* **16**:1373–1381.

Nathan, R. 2001. The challenges of studying dispersal. *Trends in Ecology & Evolution* **16**:481–483.

Nathan, R. 2006. Long-distance dispersal of plants. *Science* **313**:786–788.

Neill, W. E. 1974. The community matrix and interdependence of the competition co-efficients. *American Naturalist* **108**:399–408.

Nekola, J. C., and P. S. White. 1999. The distance decay of similarity in biogeography and ecology. *Journal of Biogeography* **26**:867–878.

Nemergut, D. R., S. K. Schmidt, T. Fukami, S. P. O'Neill, T. M. Bilinski, L. F. Stanish, J. E. Knelman, et al. 2013. Patterns and processes of microbial community assembly. *Microbiology and Molecular Biology Reviews* **77**:342–356.

Nicholson, A. J., and V. A. Bailey. 1935. The balance of animal populations. Part I. *Proceedings of the Zoological Society of London* **105**:551–598.

Noble, A., and W. Fagan. 2014. A niche remedy for the dynamical problems of neutral theory. *Theoretical Ecology* **8**:1–13.

Norberg, J. 2004. Biodiversity and ecosystem functioning: A complex adaptive systems approach. *Limnology and Oceanography* **49**:1269–1277.

Norberg, J., D. P. Swaney, J. Dushoff, J. Lin, R. Casagrandi, and S. A. Levin. 2001. Phenotypic diversity and ecosystem functioning in changing environments: A theoretical framework. *Proceedings of the National Academy of Sciences USA* **98**:11376–11381.

Norberg, J., M. C. Urban, M. Vellend, C. A. Klausmeier, and N. Loeuille. 2012. Eco-evolutionary responses of biodiversity to climate change. *Nature Climate Change* **2**:747–751.

Norden, N., S. G. Letcher, V. Boukili, N. G. Swenson, and R. Chazdon. 2011. Demographic drivers of successional changes in phylogenetic structure across life-history stages in plant communities. *Ecology* **93**:S70–S82.

Nosil, P. 2012. *Ecological speciation*. Oxford University Press, Oxford.

Nowak, M. A. 2006. *Evolutionary dynamics*. Harvard University Press, Cambridge, MA.

Odenbaugh, J. 2013. Searching for patterns, hunting for causes: Robert MacArthur, the mathematical naturalist. Pages 181–198 *in* O. Harmon and M. R. Dietrich, editors. *Outsider scientists: Routes to innovation in biology*. University of Chicago Press, Chicago.

Orr, H. A. 2009. Fitness and its role in evolutionary genetics. *Nature Reviews Genetics* **10**:531–539.

Orrock, J. L., and R. J. Fletcher Jr. 2005. Changes in community size affect the outcome of competition. *American Naturalist* **166**:107–111.

Orrock, J. L., and J. I. Watling. 2010. Local community size mediates ecological drift and competition in metacommunities. *Proceedings of the Royal Society B: Biological Sciences* **277**:2185–2191.

Otto, S. P., and T. Day. 2011. *A biologist's guide to mathematical modeling in ecology and evolution*. Princeton University Press, Princton, NJ.

Pacala, S. W., C. D. Canham, and J. Silander Jr. 1993. Forest models defined by field measurements: I. The design of a northeastern forest simulator. *Canadian Journal of Forest Research* **23**:1980–1988.

Paine, R. T. 1974. Intertidal community structure. *Oecologia* **15**:93–120.

Palmer, M. W. 1994. Variation in species richness: Towards a unification of hypotheses. *Folia Geobotanica et Phytotaxonomica* **29**:511–530.

Pandolfi, J. M., S. R. Connolly, D. J. Marshall, and A. L. Cohen. 2011. Projecting coral reef futures under global warming and ocean acidification. *Science* **333**:418–422.

Pardini, R., S. M. de Souza, R. Braga-Neto, and J. P. Metzger. 2005. The role of forest structure, fragment size and corridors in maintaining small mammal abundance and diversity in an Atlantic forest landscape. *Biological Conservation* **124**:253–266.

Parent, C. E., and B. J. Crespi. 2006. Sequential colonization and diversification of Galápagos endemic land snail genus *Bulimulus* (Gastropoda, Stylommatophora). *Evolution* **60**:2311–2328.

Park, T. 1954. Experimental studies of interspecies competition II. Temperature, humidity, and competition in two species of *Tribolium*. *Physiological Zoology* **27**:177–238.

Park, T. 1962. Beetles, competition, and populations: An intricate ecological phenomenon is brought into the laboratory and studied as an experimental model. *Science* **138**:1369–1375.

Parmesan, C. 2006. Ecological and evolutionary responses to recent climate change. *Annual Review of Ecology, Evolution, and Systematics* **37**:637–669.

Pärtel, M. 2002. Local plant diversity patterns and evolutionary history at the regional scale. *Ecology* **83**:2361–2366.

Pärtel, M., L. Laanisto, and M. Zobel. 2007. Contrasting plant productivity-diversity relationships across latitude: The role of evolutionary history. *Ecology* **88**:1091–1097.

Pärtel, M., and M. Zobel. 1999. Small-scale plant species richness in calcareous grasslands determined by the species pool, community age and shoot density. *Ecography* **22**:153–159.

Pärtel, M., M. Zobel, K. Zobel, and E. van der Maarel. 1996. The species pool and its relation to species richness: Evidence from Estonian plant communities. *Oikos* **75**:111–117.

Pedruski, M., and S. Arnott. 2011. The effects of habitat connectivity and regional heterogeneity on artificial pond metacommunities. *Oecologia* **166**:221–228.

Pelletier, F., D. Garant, and A. P. Hendry. 2009. Eco-evolutionary dynamics. *Philosophical Transactions of the Royal Society B: Biological Sciences* **364**:1483–1489.

Peters, R. H. 1991. *A critique for ecology*. Cambridge University Press, Cambridge.

Petraitis, P. 1998. How can we compare the importance of ecological processes if we never ask, "compared to what?" Pages 183–201 *in* W. J. Resetarits, Jr. and J. Bernardo, editors. *Experimental ecology: Issues and perspectives*. Oxford University Press, New York.

Pianka, E. R. 1967. On lizard species diversity: North American flatland deserts. *Ecology* **48**:334–351.

Pickett, S.T.A., S. L. Collins, and J. J. Armesto. 1987. Models, mechanisms and pathways of succession. *Botanical Review* **53**:335–371.

Pickett, S.T.A., J. Kolasa, and C. G. Jones. 2007. *Ecological understanding: The nature of theory and the theory of nature*, 2nd ed. Elsevier/Academic Press, Burlington, MA.

Pickett, S.T.A., and P. S. White. 1985. *The ecology of natural disturbance and patch dynamics*. Academic Press, San Diego.

Pigot, A. L., and R. S. Etienne. 2015. A new dynamic null model for phylogenetic community structure. *Ecology Letters* **18**:153–163.

Pimentel, D. 1968. Population regulation and genetic feedback: Evolution provides foundation for control of herbivore, parasite, and predator numbers in nature. *Science* **159**:1432–1437.

Pinto-Sánchez, N. R., A. J. Crawford, and J. J. Wiens. 2014. Using historical biogeography to test for community saturation. *Ecology Letters* **17**:1077–1085.

Platt, W. J. 1975. The colonization and formation of equilibrium plant species associations on badger disturbances in a tall-grass prairie. *Ecological Monographs* **45**:285–305.

Popper, K. 1959. *The logic of scientific discovery*. Hutchinson, London.

Prugh, L. R., K. E. Hodges, A. R. Sinclair, and J. S. Brashares. 2008. Effect of habitat area and isolation on fragmented animal populations. *Proceedings of the National Academy of Sciences USA* **105**:20770–20775.

Putnam, R. 1993. *Community ecology*. Springer, The Netherlands.

Pyron, R. A. 2014. Temperate extinction in squamate reptiles and the roots of latitudinal diversity gradients. *Global Ecology and Biogeography* **23**:1126–1134.

Pyron, R. A., and J. J. Wiens. 2013. Large-scale phylogenetic analyses reveal the causes of high tropical amphibian diversity. *Proceedings of the Royal Society B: Biological Sciences* **280**:20131622.

R Core Team. 2012. *R: A language and environment for statistical computing*. R Foundation for Statistical Computing, Vienna.

Rabosky, D. L. 2012. Testing the time-for-speciation effect in the assembly of regional biotas. *Methods in Ecology and Evolution* **3**:224–233.

Rabosky, D. L. 2013. Diversity-dependence, ecological speciation, and the role of competition in macroevolution. *Annual Review of Ecology, Evolution, and Systematics* **44**:481–502.

Ralph, C. J. 1985. Habitat association patterns of forest and steppe birds of northern Patagonia, Argentina. *Condor* **87**:471–483.

Recher, H. F. 1969. Bird species diversity and habitat diversity in Australia and North America. *American Naturalist* **103**:75–80.

Rees, M., and M. Westoby. 1997. Game-theoretical evolution of seed mass in multi-species ecological models. *Oikos* **78**:116–126.

Resetarits, W. J. Jr., and J. Bernardo. 1998. *Experimental ecology: Issues and perspectives*. Oxford University Press, New York.

Reynolds, H. L., A. Packer, J. D. Bever, and K. Clay. 2003. Grassroots ecology: Plant-microbe-soil interactions as drivers of plant community structure and dynamics. *Ecology* **84**:2281–2291.

Ricklefs, R. E. 1987. Community diversity: Relative roles of local and regional processes. *Science* **235**:167–171.

Ricklefs, R. E., and I. J. Lovette. 1999. The roles of island area per se and habitat diversity in the species–area relationships of four Lesser Antillean faunal groups. *Journal of Animal Ecology* **68**:1142–1160.

Ricklefs, R. E., and G. L. Miller. 1999. *Ecology*, 4th ed. W. H. Freeman, New York.

Ricklefs, R. E., and D. Schluter. 1993a. *Species diversity in ecological communities: Historical and geographic perspectives*. University of Chicago Press, Chicago.

Ricklefs, R. E., and D. Schluter. 1993b. Species diversity: regional and historical influences. Pages 350–363 *in* R. E. Ricklefs and D. Schluter, editors. *Species diversity in ecological communities: Historical and geographic perspectives*. University of Chicago Press, Chicago.

Ricklefs, R. E., A. E. Schwarzbach, and S. S. Renner. 2006. Rate of lineage origin explains the diversity anomaly in the world's mangrove vegetation. *American Naturalist* **168**:805–810.

Rodríguez, A., G. Jansson, and H. Andrén. 2007. Composition of an avian guild in spatially structured habitats supports a competition–colonization trade-off. *Proceedings of the Royal Society B: Biological Sciences* **274**:1403–1411.

Rodríguez, M. Á., M. Á. Olalla-Tárraga, and B. A. Hawkins. 2008. Bergmann's rule and the geography of mammal body size in the Western Hemisphere. *Global Ecology and Biogeography* **17**:274–283.

Roff, D. A. 2002. *Life history evolution*. Sinauer Associates, Sunderland, MA.

Rohde, K. 1992. Latitudinal gradients in species diversity: The search for the primary cause. *Oikos* **65**:514–527.

Rolland, J., F. L. Condamine, F. Jiguet, and H. Morlon. 2014. Faster speciation and reduced extinction in the tropics contribute to the mammalian latitudinal diversity gradient. *PLoS Biology* **12**:e1001775.

Root, R. B. 1967. The niche exploitation pattern of the blue-gray gnatcatcher. *Ecological Monographs* **37**:317–350.

Rosenblum, E. B., B. A. Sarver, J. W. Brown, S. Des Roches, K. M. Hardwick, T. D. Hether, J. M. Eastman, et al. 2012. Goldilocks meets Santa Rosalia: An ephemeral speciation model explains patterns of diversification across time scales. *Evolutionary Biology* **39**:255–261.

Rosenzweig, M. L. 1975. On continental steady states of species diversity. Pages 121–140 *in* M. L. Cody and J. M. Diamond, editors. *Ecology and evolution of communities*. Belknap Press of Harvard University Press, Cambridge, MA.

Rosenzweig, M. L. 1995. *Species diversity in space and time*. Cambridge University Press, Cambridge.

Rosindell, J., S. J. Cornell, S. P. Hubbell, and R. S. Etienne. 2010. Protracted speciation revitalizes the neutral theory of biodiversity. *Ecology Letters* **13**:716–727.

Rosindell, J., S. P. Hubbell, and R. S. Etienne. 2011. The unified neutral theory of bio-diversity and biogeography at age ten. *Trends in Ecology & Evolution* **26**:340–348.

Rosindell, J., S. P. Hubbell, F. He, L. J. Harmon, and R. S. Etienne. 2012. The case for ecological neutral theory. *Trends in Ecology & Evolution* **27**:203–208.

Rosindell, J., and A. B. Phillimore. 2011. A unified model of island biogeography sheds light on the zone of radiation. *Ecology Letters* **14**:552–560.

Roughgarden, J. 2009. Is there a general theory of community ecology? *Biology & Philosophy* **24**:521–529.

Roxburgh, S. H., K. Shea, and J. B. Wilson. 2004. The intermediate disturbance hypoth-esis: Patch dynamics and mechanisms of species coexistence. *Ecology* **85**:359–371.

Roy, K., J. W. Valentine, D. Jablonski, and S. M. Kidwell. 1996. Scales of climatic variability and time averaging in Pleistocene biotas: Implications for ecology and evolution. *Trends in Ecology & Evolution* **11**:458–463.

Rull, V. 2013. Some problems in the study of the origin of neotropical biodiversity using palaeoecological and molecular phylogenetic evidence. *Systematics and Biodiversity* **11**:415–423.

Rundle, H. D., and P. Nosil. 2005. Ecological speciation. *Ecology Letters* **8**:336–352.

Sagarin, R., and A. Pauchard. 2012. *Observation and ecology: Broadening the scope of science to understand a complex world*. Island Press, Washington, DC.

Sattler, T., D. Borcard, R. Arlettaz, F. Bontadina, P. Legendre, M. K. Obrist, and M. Moretti. 2010. Spider, bee, and bird communities in cities are shaped by environmental con-trol and high stochasticity. *Ecology* **91**:3343–3353.

Sax, D. F., and S. D. Gaines. 2003. Species diversity: from global decreases to local increases. *Trends in Ecology & Evolution* **18**:561–566.

Sax, D. F., J. J. Stachowicz, J. H. Brown, J. F. Bruno, M. N. Dawson, S. D. Gaines, R. K. Gros-berg, et al. 2007. Ecological and evolutionary insights from species invasions. *Trends in Ecology & Evolution* **22**:465–471.

Scheffer, M. 2009. *Critical transitions in nature and society*. Princeton University Press, Princeton, N.J.

Scheffer, M., S. Carpenter, J. A. Foley, C. Folke, and B. Walker. 2001. Catastrophic shifts in ecosystems. *Nature* **413**:591–596.

Scheffer, M., and S. R. Carpenter. 2003. Catastrophic regime shifts in ecosystems: Link-ing theory to observation. *Trends in Ecology & Evolution* **18**:648–656.

Scheffer, M., and E. H. van Nes. 2006. Self-organized similarity, the evolutionary emer-gence of groups of similar species. *Proceedings of the National Academy of Sciences USA* **103**:6230–6235.

Scheffer, M., S. Hosper, M. Meijer, B. Moss, and E. Jeppesen. 1993. Alternative equi-libria in shallow lakes. *Trends in Ecology & Evolution* **8**:275–279.

Scheiner, S. M., and M. R. Willig. 2011. *The theory of ecology*. University of Chicago Press, Chicago.

Schluter, D. 2000. *The ecology of adaptive radiation*. Oxford University Press, Oxford.

Schluter, D., and R. E. Ricklefs. 1993. Convergence and the regional component of species diversity. Pages 230–240 *in* R. E. Ricklefs and D. Schluter, editors. *Species diversity in ecological communities: Historical and geographic perspectives.* University of Chicago Press, Chicago.

Schoener, T. W. 1974. Resource partitioning in ecological communities. *Science* **185**: 27–39.

Schoener, T. W. 1983a. Field experiments on interspecific competition. *American Naturalist* **122**:240–285.

Schoener, T. W. 1983b. Rate of species turnover decreases from lower to higher organisms: A review of the data. *Oikos* **41**:372–377.

Schoener, T. W. 2011. The newest synthesis: Understanding the interplay of evolutionary and ecological dynamics. *Science* **331**:426–429.

Seabloom, E. W., E. T. Borer, K. Gross, A. E. Kendig, C. Lacroix, C. E. Mitchell, E. A. Mordecai, et al. 2015. The community ecology of pathogens: Coinfection, coexistence and community composition. *Ecology Letters* **18**:401–415.

Seiferling, I., R. Proulx, and C. Wirth. 2014. Disentangling the environmental-heterogeneity–species-diversity relationship along a gradient of human footprint. *Ecology* **95**:2084–2095.

Shipley, B. 2002. *Cause and correlation in biology: A user's guide to path analysis, structural equations and causal inference.* Cambridge University Press, Cambridge.

Shipley, B. 2010. *From plant traits to vegetation structure: chance and selection in the assembly of ecological communities.* Cambridge University Press, Cambridge.

Shipley, B., D. Vile, and É. Garnier. 2006. From plant traits to plant communities: A statistical mechanistic approach to biodiversity. *Science* **314**:812–814.

Shmida, A., and M. V. Wilson. 1985. Biological determinants of species diversity. *Journal of Biogeography* **12**:1–20.

Shrader-Frechette, K. S. and D. McCoy. 1993. *Method in ecology: Strategies for conservation.* Cambridge University Press, Cambridge.

Shurin, J., and D. S. Srivastava. 2005. New perspectives on local and regional diversity: Beyond saturation. Pages 399–417 *in* M. Holyoak, R. D. Holt, and M. A. Leibold, editors. *Metacommunities.* University of Chicago Press, Chicago.

Shurin, J. B. 2000. Dispersal limitation, invasion resistance, and the structure of pond zooplankton communities. *Ecology* **81**:3074–3086.

Shurin, J. B., E. T. Borer, E. W. Seabloom, K. Anderson, C. A. Blanchette, B. Broitman, S. D. Cooper, et al. 2002. A cross-ecosystem comparison of the strength of trophic cascades. *Ecology Letters* **5**:785–791.

Shurin, J. B., K. Cottenie, and H. Hillebrand. 2009. Spatial autocorrelation and dispersal limitation in freshwater organisms. *Oecologia* **159**:151–159.

Siepielski, A. M., K.-L. Hung, E. E. B. Bein, and M. A. McPeek. 2010. Experimental evidence for neutral community dynamics governing an insect assemblage. *Ecology* **91**:847–857.

Siepielski, A. M,. and M. A. McPeek. 2010. On the evidence for species coexistence: A critique of the coexistence program. *Ecology* **91**:3153–3164.

Siepielski, A. M., and M. A. McPeek. 2013. Niche versus neutrality in structuring the beta diversity of damselfly assemblages. *Freshwater Biology* **58**:758–768.

Simberloff, D. 2004. Community ecology: Is it time to move on? *American Naturalist* **163**:787–799.

Simberloff, D., and B. Von Holle. 1999. Positive interactions of nonindigenous species: Invasional meltdown? *Biological Invasions* **1**:21–32.

Simonis, J. L., and J. C. Ellis. 2013. Bathing birds bias β-diversity: Frequent dispersal by gulls homogenizes fauna in a rock-pool metacommunity. *Ecology* **95**:1545–1555.

Sinervo, B., and C. M. Lively. 1996. The rock-paper-scissors game and the evolution of alternative male strategies. *Nature* **380**:240–243.

Slatkin, M. 1974. Competition and regional coexistence. *Ecology* **55**:128–134.

Smol, J. P., and E. F. Stoermer. 2010. *The diatoms: Applications for the environmental and earth sciences.* Cambridge University Press, Cambridge.

Sober, E. 1991. Models of cultural evolution. Pages 17–38 *in* P. Griffiths, editor. *Trees of life: Essays in the philosophy of biology.* Kluwer, New York.

Sober, E. 2000. *Philosophy of biology*, 2nd ed. Westview Press, Boulder, CO.

Soininen, J. 2014. A quantitative analysis of species sorting across organisms and ecosystems. *Ecology* **95**:3284–3292.

Soininen, J., R. McDonald, and H. Hillebrand. 2007. The distance decay of similarity in ecological communities. *Ecography* **30**:3–12.

Sommer, B., P. L. Harrison, M. Beger, and J. M. Pandolfi. 2013. Trait-mediated environmental filtering drives assembly at biogeographic transition zones. *Ecology* **95**:1000–1009.

Srivastava, D. S. 1999. Using local–regional richness plots to test for species saturation: Pitfalls and potentials. *Journal of Animal Ecology* **68**:1–16.

Stanley, S. M. 1979. *Macroevolution, pattern and process.* Johns Hopkins University Press, Baltimore.

Stauffer, R. C. 1957. Haeckel, Darwin, and ecology. *Quarterly Review of Biology* **32**:138–144.

Staver, A. C., S. Archibald, and S. A. Levin. 2011. The global extent and determinants of savanna and forest as alternative biome states. *Science* **334**:230–232.

Stein, A., K. Gerstner, and H. Kreft. 2014. Environmental heterogeneity as a universal driver of species richness across taxa, biomes and spatial scales. *Ecology Letters* **17**:866–880.

Stephens, P. R., and J. J. Wiens. 2003. Explaining species richness from continents to communities: The time-for-speciation effect in emydid turtles. *American Naturalist* **161**:112–128.

Stevens, H. 2009. *A primer of ecology with R.* Springer, New York.

Stockwell, C. A., A. P. Hendry, and M. T. Kinnison. 2003. Contemporary evolution meets conservation biology. *Trends in Ecology & Evolution* **18**:94–101.

Strong, D. R., D. Simberloff, L. G. Abele, and A. B. Thistle. 1984. *Ecological communities: Conceptual issues and the evidence.* Princeton University Press, Princeton, NJ.

Suding, K. N., K. L. Gross, and G. R. Houseman. 2004. Alternative states and positive feedbacks in restoration ecology. *Trends in Ecology & Evolution* **19**:46–53.

Szava-Kovats, R. C., A. Ronk, and M. Pärtel. 2013. Pattern without bias: Local–regional richness relationship revisited. *Ecology* **94**:1986–1992.

Tamme, R., I. Hiiesalu, L. Laanisto, R. Szava-Kovats, and M. Pärtel. 2010. Environmental heterogeneity, species diversity and co-existence at different spatial scales. *Journal of Vegetation Science* **21**:796–801.

Tansley, A. 1917. On competition between *Galium saxatile* L. (*G. hercynicum* Weig.) and *Galium sylvestre* Poll. (*G. asperum* Schreb.) on different types of soil. *Journal of Ecology* **5**:173–179.

Tansley, A. G. 1939. *The British Isles and their vegetation.* Cambridge University Press, Cambridge.

Taylor, D. R., L. W. Aarssen, and C. Loehle. 1990. On the relationship between r/K selection and environmental carrying capacity: A new habitat templet for plant life history strategies. *Oikos* **58**:239–250.

Terborgh, J. W., and J. Faaborg. 1980. Saturation of bird communities in the West Indies. *American Naturalist* **116**:178–195.

Tews, J., U. Brose, V. Grimm, K. Tielbörger, M. C. Wichmann, M. Schwager, and F. Jeltsch. 2004. Animal species diversity driven by habitat heterogeneity/diversity: The importance of keystone structures. *Journal of Biogeography* **31**:79–92.

Tilman, D. 1977. Resource competition between plankton algae: An experimental and theoretical approach. *Ecology* **58**:338–348.

Tilman, D. 1981. Tests of resource competition theory using four species of Lake Michigan algae. *Ecology* **62**:802–815.

Tilman, D. 1982. *Resource competition and community structure.* Princeton University Press, Princeton, NJ.

Tilman, D. 1994. Competition and biodiversity in spatially structured habitats. *Ecology* **75**:2–16.

Tilman, D. 1997. Community invasibility, recruitment limitation, and grassland biodiversity. *Ecology* **78**:81–92.

Tilman, D. 2004. Niche tradeoffs, neutrality, and community structure: A stochastic theory of resource competition, invasion, and community assembly. *Proceedings of the National Academy of Sciences USA* **101**:10854–10861.

Tilman, D. 2011. Diversification, biotic interchange, and the universal trade-off hypothesis. *American Naturalist* **178**:355–371.

Tilman, D. and P. M. Kareiva. 1997. *Spatial ecology: The role of space in population dynamics and interspecific interactions.* Princeton University Press, Princeton, NJ.

Tilman, D., S. S. Kilham, and P. Kilham. 1982. Phytoplankton community ecology: The role of limiting nutrients. *Annual Review of Ecology and Systematics* **13**:349–372.

Tokeshi, M. 1999. *Species coexistence: Ecological and evolutionary perspectives.* Wiley, Oxford.

Tucker, C., and M. Cadotte. 2013. Reinventing the wheel—why do we do it? *The EEB & Flow.* evol-eco.blogspot.ca/2013/01/reinventing-ecological-wheel-why-do-we.html.

Tucker, C. M., and T. Fukami. 2014. Environmental variability counteracts priority effects to facilitate species coexistence: Evidence from nectar microbes. *Proceedings of the Royal Society B: Biological Sciences* **281**:20132637.

Tuomisto, H., and K. Ruokolainen. 2006. Analyzing or explaining beta diversity? Understanding the targets of different methods of analysis. *Ecology* **87**:2697–2708.

Tuomisto, H., K. Ruokolainen, and M. Yli-Halla. 2003. Dispersal, environment, and floristic variation of western Amazonian forests. *Science* **299**:241–244.

Turgeon, J., R. Stoks, R. A. Thum, J. M. Brown, and M. A. McPeek. 2005. Simultaneous Quaternary radiations of three damselfly clades across the Holarctic. *American Naturalist* **165**:E78–E107.

Turnbull, L. A., M. J. Crawley, and M. Rees. 2000. Are plant populations seed-limited? a review of seed sowing experiments. *Oikos* **88**:225–238.

Turnbull, L. A., J. M. Levine, M. Loreau, and A. Hector. 2013. Coexistence, niches and biodiversity effects on ecosystem functioning. *Ecology Letters* **16**:116–127.

Turnbull, L. A., M. Rees, and M. J. Crawley. 1999. Seed mass and the competition/colonization trade-off: A sowing experiment. *Journal of Ecology* **87**:899–912.

Urban, M. C., M. A. Leibold, P. Amarasekare, L. De Meester, R. Gomulkiewicz, M. E. Hochberg, C. A. Klausmeier, et al. 2008. The evolutionary ecology of metacommunities. *Trends in Ecology & Evolution* **23**:311–317.

USGS. 2013. North American Breeding Bird Survey FTP data set. Version 2013.0. USGS Patuxent Wildlife Research Center, Laurel, MD.

Vamosi, S., S. Heard, J. Vamosi, and C. Webb. 2009. Emerging patterns in the comparative analysis of phylogenetic community structure. *Molecular Ecology* **18**:572–592.

van der Plas, F., T. Janzen, A. Ordonez, W. Fokkema, J. Reinders, R. S. Etienne, and H. Olff. 2015. A new modeling approach estimates the relative importance of different community assembly processes. *Ecology* **96**:1502–1515.

van der Valk, A. 2011. Origins and development of ecology. Pages 25–48 *in* K. deLaplante, B. Brown, and K. A. Peacock, editors. *Philosophy of ecology*. Elsevier, Oxford.

Van Geest, G. J., F.C.J.M. Roozen, H. Coops, R.M M. Roijackers, A. D. Buijse, E.T.H.M. Peeters, and M. Scheffer. 2003. Vegetation abundance in lowland flood plan lakes determined by surface area, age and connectivity. *Freshwater Biology* **48**:440–454.

Van Valen, L., and F. A. Pitelka. 1974. Intellectual censorship in ecology. *Ecology* **55**:925–926.

Vandermeer, J. H. 1969. The competitive structure of communities: An experimental approach with protozoa. *Ecology* **50**:362–371.

Vanschoenwinkel, B., F. Buschke, and L. Brendonck. 2013. Disturbance regime alters the impact of dispersal on alpha and beta diversity in a natural metacommunity. *Ecology* **94**:2547–2557.

Vázquez-Rivera, H., and D. J. Currie. 2015. Contemporaneous climate directly controls broad-scale patterns of woody plant diversity: A test by a natural experiment over 14,000 years. *Global Ecology and Biogeography* **24**:97–106.

Vellend, M. 2004. Parallel effects of land-use history on species diversity and genetic diversity of forest herbs. *Ecology* **85**:3043–3055.

Vellend, M. 2006. The consequences of genetic diversity in competitive communities. *Ecology* **87**:304–311.

Vellend, M. 2010. Conceptual synthesis in community ecology. *Quarterly Review of Biology* **85**:183–206.

Vellend, M., L. Baeten, I. H. Myers-Smith, S. C. Elmendorf, R. Beauséjour, C. D. Brown, P. De Frenne, et al. 2013. Global meta-analysis reveals no net change in local-scale plant biodiversity over time. *Proceedings of the National Academy of Sciences USA* **110**:19456–19459.

Vellend, M., W. K. Cornwell, K. Magnuson-Ford, and A. O. Mooers. 2010. Measuring phylogenetic biodiversity. Pages 193–206 *in* A. E. Magurran and B. McGill, editors. *Biological diversity: Frontiers in measurement and assessment*. Oxford University Press, New York.

Vellend, M., and M. A. Geber. 2005. Connections between species diversity and genetic diversity. *Ecology Letters* **8**:767–781.

Vellend, M., and I. Litrico. 2008. Sex and space destabilize intransitive competition within and between species. *Proceedings of the Royal Society B: Biological Sciences* **275**:1857–1864.

Vellend, M., J. A. Myers, S. Gardescu, and P. Marks. 2003. Dispersal of *Trillium* seeds by deer: Implications for long-distance migration of forest herbs. *Ecology* **84**:1067–1072.

Vellend, M., and J. L. Orrock. 2009. Genetic and ecological models of diversity: Lessons across disciplines. Pages 439–461 *in* J. B. Losos and R. E. Ricklefs, editors. *The theory of island biogeography revisited*. Princeton University Press, Princeton, NJ.

Vellend, M., D. S. Srivastava, K. M. Anderson, C. D. Brown, J. E. Jankowski, E. J. Kleynhans, N. J. B. Kraft, et al. 2014. Assessing the relative importance of neutral stochasticity in ecological communities. *Oikos* **123**:1420–1430.

Vellend, M., K. Verheyen, K. M. Flinn, H. Jacquemyn, A. Kolb, H. Van Calster, G. Peterken, B. J., et al. 2007. Homogenization of forest plant communities and weakening of species–environment relationships via agricultural land use. *Journal of Ecology* **95**:565–573.

Vellend, M., K. Verheyen, H. Jacquemyn, A. Kolb, H. Van Calster, G. Peterken, and M. Hermy. 2006. Extinction debt of forest plants persists for more than a century following habitat fragmentation. *Ecology* **87**:542–548.

Verheyen, K., O. Honnay, G. Motzkin, M. Hermy, and D. R. Foster. 2003. Response of forest plant species to land-use change: A life-history trait-based approach. *Journal of Ecology* **91**:563–577.

Verhoef, H. A., and P. J. Morin, eds. 2010. *Community ecology: Processes, models, and applications*. Oxford University Press, New York.

Vermeij, G. J. 1991. When biotas meet: Understanding biotic interchange. *Science* **253**:1099–1104.

Vermeij, G. J. 2005. Invasion as expectation: A historical fact of life. Pages 315–339 *in* D. F. Sax, J. J. Stachowicz, and S. D. Gaines, editors. *Species invasions: Insights into ecology, evolution, and biogeography*. Sinauer Associates, Sunderland, MA.

Violle, C., M. L. Navas, D. Vile, E. Kazakou, C. Fortunel, I. Hummel, and E. Garnier. 2007. Let the concept of trait be functional! *Oikos* **116**:882–892.

Wagner, C. E., L. J. Harmon, and O. Seehausen. 2014. Cichlid species-area relationships are shaped by adaptive radiations that scale with area. *Ecology Letters* **17**:583–592.

Waide, R. B., M. R. Willig, C. F. Steiner, G. Mittelbach, L. Gough, S. I. Dodson, G. P. Juday, et al. 1999. The relationship between productivity and species richness. *Annual Review of Ecology and Systematics* **30**:257–300.

Wardle, D. A., O. Zackrisson, G. Hörnberg, and C. Gallet. 1997. The influence of island area on ecosystem properties. *Science* **277**:1296–1299.

Warren, P. H. 1996. Dispersal and destruction in a multiple habitat system: An experimental approach using protist communities. *Oikos* **77**:317–325.

Warton, D. I., S. D. Foster, G. De'ath, J. Stoklosa, and P. K. Dunstan. 2015. Model-based thinking for community ecology. *Plant Ecology* 216:669–682.

Webb, C. O., D. D. Ackerly, M. A. McPeek, and M. J. Donoghue. 2002. Phylogenies and community ecology. *Annual Review of Ecology and Systematics* **33**:475–505.

Weiher, E. 2010. A primer of trait and functional diversity. Pages 175–193 *in* A. E. Magurran and B. J. McGill, editors. *Biological diversity: Frontiers in measurement and assessment*. Oxford University Press, New York.

Weiher, E., D. Freund, T. Bunton, A. Stefanski, T. Lee, and S. Bentivenga. 2011. Advances, challenges and a developing synthesis of ecological community assembly theory. *Philosophical Transactions of the Royal Society B: Biological Sciences* **366**:2403–2413.

Weiher, E., and P. A. Keddy. 1995. Assembly rules, null models, and trait dispersion: New questions from old patterns. *Oikos* **74**:159–164.

Weiher, E., and P. A. Keddy. 2001. *Ecological assembly rules: Perspectives, advances, retreats*. Cambridge University Press, Cambridge.

Weir, J. T., and D. Schluter. 2007. The latitudinal gradient in recent speciation and extinction rates of birds and mammals. *Science* **315**:1574–1576.

Werner, E. E. 1998. Ecological experiments and a research program in community ecology. Pages 3–26 *in* W. J. Resetarits, Jr. and J. Bernardo, editors. *Experimental ecology: Issues and perspectives*. Oxford University Press, New York.

White, E. P., and A. H. Hurlbert. 2010. The combined influence of the local environment and regional enrichment on bird species richness. *American Naturalist* **175**:E35–E43.

Whittaker, R. H. 1956. Vegetation of the Great Smoky Mountains. *Ecological Monographs* **26**:1–80.

Whittaker, R. H. 1960. Vegetation of the Siskiyou Mountains, Oregon and California. *Ecological Monographs* **30**:279–338.

Whittaker, R. H. 1975. *Communities and ecosystems*. Macmillan, New York.

Whittaker, R. J., and J. M. Fernandez-Palacios. 2007. *Island biogeography: Ecology, evolution, and conservation*. Oxford University Press, Oxford.

Wiens, J. J. 2011. The causes of species richness patterns across space, time, and clades and the role of "ecological limits." *Quarterly Review of Biology* **86**:75–96.

Wiens, J. J., and M. J. Donoghue. 2004. Historical biogeography, ecology and species richness. *Trends in Ecology & Evolution* **19**:639–644.

Wiens, J. J., G. Parra-Olea, M. García-París, and D. B. Wake. 2007. Phylogenetic history underlies elevational biodiversity patterns in tropical salamanders. *Proceedings of the Royal Society B: Biological Sciences* **274**:919–928.

Wiens, J. J., R. A. Pyron, and D. S. Moen. 2011. Phylogenetic origins of local-scale diversity patterns and the causes of Amazonian megadiversity. *Ecology Letters* **14**:643–652.

Williams, J. W, and S. T. Jackson. 2007. Novel climates, no-analog communities, and ecological surprises. *Frontiers in Ecology and the Environment* **5**:475–482.

Williamson, M. 1988. Relationship of species number to area, distance and other variables. Pages 91–115 *in* A. A. Myers and P. S. Giller, editors. *Analytical biogeography: An integrated approach to the study of animal and plant distributions*. Chapman & Hall, London.

Wilson, E. O. 2013. *Letters to a young scientist*. Liveright, New York.

Wilson, J. B., and A.D.Q. Agnew. 1992. Positive-feedback switches in plant communities. *Advances in Ecological Research* **23**:263–336.

Wolkovich, E. M., B. I. Cook, J. M. Allen, T. M. Crimmins, J. L. Betancourt, S. E. Travers, S. Pau, et al. 2012. Warming experiments underpredict plant phenological responses to climate change. *Nature* **485**:494–497.

Worster, D. 1994. *Nature's economy: A history of ecological ideas*. Cambridge University Press, Cambridge.

Wright, D. H. 1983. Species-energy theory: An extension of species-area theory. *Oikos* **41**:496–506.

Wright, I. J., P. B. Reich, M. Westoby, D. D. Ackerly, Z. Baruch, F. Bongers, et al. 2004. The worldwide leaf economics spectrum. *Nature* **428**:821–827.

Wright, S. 1940. Breeding structure of populations in relation to speciation. *American Naturalist* **74**:232–248.

Wright, S. 1964. Biology and the philosophy of science. *Monist* **48**:265–290.

Wright, S. D., R. D. Gray, and R. C. Gardner. 2003. Energy and the rate of evolution: Inferences from plant rDNA substitution rates in the western Pacific. *Evolution* **57**:2893–2898.

Wright, S. J., K. Kitajima, N.J.B. Kraft, P. B. Reich, I. J. Wright, D. E. Bunker, R. Condit, et al. 2010. Functional traits and the growth–mortality trade-off in tropical trees. *Ecology* **91**:3664–3674.

Yawata, Y., O. X. Cordero, F. Menolascina, J.-H. Hehemann, M. F. Polz, and R. Stocker. 2014. Competition–dispersal tradeoff ecologically differentiates recently speciated marine bacterioplankton populations. *Proceedings of the National Academy of Sciences USA* **111**:5622–5627.

Yeaton, R., and W. Bond. 1991. Competition between two shrub species: Dispersal differences and fire promote coexistence. *American Naturalist* **138**:328–341.

Yi, X., and A. M. Dean. 2013. Bounded population sizes, fluctuating selection and the tempo and mode of coexistence. *Proceedings of the National Academy of Sciences USA* **110**:16945–16950.

Yodzis, P. 1988. The indeterminacy of ecological interactions as perceived through perturbation experiments. *Ecology* **69**:508–515.

Yu, D. W., and H. B. Wilson. 2001. The competition-colonization trade-off is dead; long live the competition-colonization trade-off. *American Naturalist* **158**:49–63.

Zobel, M. 1997. The relative of species pools in determining plant species richness: An alternative explanation of species coexistence? *Trends in Ecology & Evolution* **12**:266–269.

Index

MONOGRAPHS IN POPULATION BIOLOGY

EDITED BY SIMON A. LEVIN AND HENRY S. HORN